Predicted and Measured Behavior of Five Spread Footings On Sand

Proceedings of a Prediction Symposium Sponsored by the Federal Highway Administration at the occasion of the Settlement ´94 ASCE Conference at Texas A&M University, June 16-18, 1994.

Edited by Jean-Louis Briaud and Robert M. Gibbens

Geotechnical Special Publication No. 41

Published by the
American Society of Civil Engineers
345 East 47th Street
New York, New York 10017-2398

ABSTRACT

This publication contains the results of a full scale load testing program used to predict and measure the behavior of five spread footings on sand. These results were presented at the Prediction Symposium sponsored by the Federal Highway Administration at the occasion of the Settlement '94 ASCE Conference at Texas A&M University, June 16-18, 1994. At Texas A&M University National Geotechnical Experimentation Site five spread footings ranging in size from 1x1 m to 3x3 m were load tested to 150 mm of displacement on sandy soil. These papers present the results of the load tests; present the data upon which all predictions are based; describe methodologies and judgments used to make the predictions; and summarize the predictions and compare predicted responses.

Library of Congress Cataloging-in-Publication Data

Predicted and measured behavior of five spread footings on sand: proceedings of a prediction symposium / sponsored by the Federal Highway Administration at the occasion of the settlement '94 ASCE conference at Texas A&M University, June 16-18, 1994; edited by Jean-Louis Briaud and Robert Gibbens.

p. cm. — (Geotechnical special publication; no. 41)
Includes indexes.
ISBN 0-7844-0025-3
1. Bridges—Foundations and piers—Testing. 2. Concrete footings-Testing. 3. Sand—Testing. I. Briaud, J.-L. II. Gibbens, Robert (Robert M.) III. United States. Federal Highway Administration. IV. American Society of Civil Engineers. V. Series.
TG320.P74 1994 94-19960
624'.254—dc20 CIP

Library of Congress Catalog Card No: 94-19960
ISBN 0-7844-0025-3
Manufactured in the United States of America.

GEOTECHNICAL SPECIAL PUBLICATIONS

1) TERZAGHI LECTURES
2) GEOTECHNICAL ASPECTS OF STIFF AND HARD CLAYS
3) LANDSLIDE DAMS: PROCESSES RISK, AND MITIGATION
4) TIEBACKS FOR BULKHEADS
5) SETTLEMENT OF SHALLOW FOUNDATION ON COHESIONLESS SOILS: DESIGN AND PERFORMANCE
6) USE OF IN SITU TESTS IN GEOTECHNICAL ENGINEERING
7) TIMBER BULKHEADS
8) FOUNDATIONS FOR TRANSMISSION LINE TOWERS
9) FOUNDATIONS AND EXCAVATIONS IN DECOMPOSED ROCK OF THE PIEDMONT PROVINCE
10) ENGINEERING ASPECTS OF SOIL EROSION DISPERSIVE CLAYS AND LOESS
11) DYNAMIC RESPONSE OF PILE FOUNDATIONS— EXPERIMENT, ANALYSIS AND OBSERVATION
12) SOIL IMPROVEMENT - A TEN YEAR UPDATE
13) GEOTECHNICAL PRACTICE FOR SOLID WASTE DISPOSAL '87
14) GEOTECHNICAL ASPECTS OF KARST TERRIANS
15) MEASURED PERFORMANCE SHALLOW FOUNDATIONS
16) SPECIAL TOPICS IN FOUNDATIONS
17) SOIL PROPERTIES EVALUATION FROM CENTRIFUGAL MODELS
18) GEOSYNTHETICS FOR SOIL IMPROVEMENT
19) MINE INDUCED SUBSIDENCE: EFFECTS ON ENGINEERED STRUCTURES
20) EARTHQUAKE ENGINEERING & SOIL DYNAMICS (II)
21) HYDRAULIC FILL STRUCTURES
22) FOUNDATION ENGINEERING
23) PREDICTED AND OBSERVED AXIAL BEHAVIOR OF PILES
24) RESILIENT MODULI OF SOILS: LABORATORY CONDITIONS
25) DESIGN AND PERFORMANCE OF EARTH RETAINING STRUCTURES
26) WASTE CONTAINMENT SYSTEMS; CONSTRUCTION, REGULATION, AND PERFORMANCE
27) GEOTECHNICAL ENGINEERING CONGRESS
28) DETECTION OF AND CONSTRUCTION AT THE SOIL/ROCK INTERFACE
29) RECENT ADVANCES IN INSTRUMENTATION, DATA ACQUISITION AND TESTING IN SOIL DYNAMICS
30) GROUTING, SOIL IMPROVEMENT AND GEOSYNTHETICS
31) STABILITY AND PERFORMANCE OF SLOPES AND EMBANKMENTS II (A 25-YEAR PERSPECTIVE)
32) EMBANKMENT DAMS-JAMES L. SHERARD CONTRIBUTIONS
33) EXCAVATION AND SUPPORT FOR THE URBAN INFRASTRUCTURE
34) PILES UNDER DYNAMIC LOADS
35) GEOTECHNICAL PRACTICE IN DAM REHABILITATION
36) FLY ASH FOR SOIL IMPROVEMENT
37) ADVANCES IN SITE CHARACTERIZATION: DATA ACQUISITION, DATA MANAGEMENT AND DATA INTERPRETATION
38) DESIGN AND PERFORMANCE OF DEEP FOUNDATIONS: PILES AND PIERS IN SOIL AND SOFT ROCK
39) UNSATURATED SOILS
40) VERTICAL AND HORIZONTAL DEFORMATIONS OF FOUNDATIONS AND EMBANKMENTS
41) PREDICTED AND MEASURED BEHAVIOR OF FIVE SPREAD FOOTINGS ON SAND

CONTENTS

PREDICTION PAPERS

PREFACE AND ACKNOWLEDGMENTS

The Federal Highway Administration is attempting to increase industry confidence in the use of spread footings by performing quality spread footing research (see Al Dimillio's Introduction). In 1991, the vision of Mr. Al Dimillio of the Federal Highway Administration and Dr. Jean-Louis Briaud of Texas A&M University began to take shape with the drafting of a $650,000, six phase spread footing research project. The contract was between FHWA, Geotest Engineering and Texas A&M University. The project was to center around two main themes: a full scale load testing program including a comprehensive series of field and laboratory tests and the development of a spread footing database program. The load and field/laboratory testing programs were performed at both the National Geotechnical Experimentation Site on the Riverside Campus of Texas A&M University and the main Texas A&M University campus.

It was felt that a prediction symposium would be the best method to evaluate current industry and academic practice in the field of spread footing design. Due to good timing and a keen interest, the prediction symposium was scheduled to coincide with the ASCE specialty conference Settlement 94' on the campus of Texas A&M University in June 1994. More than 140 people from all over the world expressed interest in receiving information on the prediction symposium. In July 1993, a prediction package which included the results of all field tests, a description of the footing dimensions and layout and a prediction request was sent to those interested in making a prediction. This prediction package and addendums are presented by Gibbens and Briaud in this volume. The following field and laboratory tests were included in the prediction package:

TEST	DATE
Borehole Shear Test	April 18 & 19, 1993
Cross-Hole Wave Tests	May 13, 1993
PiezoCone Penetration Test	January 27, 1993
Dilatometer Test	April 8 & 13, 1993

Pressuremeter Tests	February 2, 1993 May 15, 1993
Step Blade Test	April 19, 1993
Standard Penetration Tests	April 5-21, 1993
Water Content & Unit Weights	November 1992-May 1993
Atterberg Limits	May 1993
Relative Density	November 1992
Triaxial Tests	April 1993
Resonant Column Tests	May 1993

The results of these tests were compiled into the prediction package both graphically and in tabular form. Beginning in September 1993 and running through March 1994, a total of 31 predictions were received including 16 from academics and 15 from consultants. Of the 31 predictors, 21 were from the United States and 10 were from foreign countries.

Load tests began in November 1993 and were completed in December 1993. Five spread footings ranging in size from lxl m to 3x3 m were load tested to 150 mm of displacement on sandy soil at the Texas A&M University National Geotechnical Experimentation Site. The results of the load test and of the predictions are in the paper by Briaud and Gibbens. All the prediction papers follow.

All the predictors are thanked for their participation in the prediction event. They made it possible for our profession to evaluate our current practice and should pride themselves for this important professional contribution.

The following persons and companies are also thanked for their contributions to this project:

This project is sponsored by the Federal Highway Administration in Washington. Mr. Al DiMillio and Mr. Michael Adams are the technical contacts for this project at FHWA. They both went out of their way to help make this project a success. The project is being performed by Geotest Engineering in cooperation with Texas A&M University. Dr. V.N. Vijayvergiya is in charge of the project at Geotest Engineering, Inc., while Dr. Jean-Louis Briaud is in charge at Texas A&M University.

Dr. Silvano Marchetti sent Mr. Maurizio Calabrese of the University of L'Aquila all the way from Italy to perform the Dilatometer Tests free of charge. Both are considered to be cosponsors of the project.

Dr. Phillip Buchanan and his team at Buchanan/Soil Mechanics, Inc (B/SMI) in Bryan, Texas provided, at a reduced rate, all the drilling rig services including those for Standard Penetration Tests, for Pressuremeter Tests, for Dilatometer Tests and for setting casings for Cross Hole Tests, Inclinometer Tests and Telltales; B/SMI is considered to be a cosponsor of the project.

Dr. George Goble of Goble Rausche Likins and Associates, Inc (GRL) came from Boulder, CO to perform the SPT energy measurements and charged a very reduced rate; GRL is considered to be a cosponsor of the project.

Mr. Recep Yilmaz and his team at Fugro McClelland performed the PiezoCone Penetrometer Tests.

Dr. Allan Lutenegger and Dr. Don Degroot of the University of Massachusetts performed the Step Blade Test, the Borehole Shear Tests and provided the load cell for the axial thrust measurements during the Dilatometer Test. Mr. Michael Adams of FHWA performed the Borehole Shear Test along with Dr. Lutenegger and Dr. Degroot.

Dr. Ken Stokoe of the University of Texas provided the second Dilatometer apparatus used for the DMT sounding, performed by Mr. Robert Gibbens independently of Mr. Calabrese's tests.

Dr. Derek Morris and Mr. Tony Yen of Texas A&M University performed the Resonant Column Tests and the Cross Hole Tests.

Mr. Phillipe Jeanjean of Texas A&M University performed the Triaxial Tests and the Pressuremeter Tests with Dr. Jean-Louis Briaud.

Mr. Glyen Farmer of Drillers Inc. in Houston, Texas directed the construction of the footings, the reaction shafts and the load test frame program. His team operated the jack during the load test. The project would not have been as successful as it was without their help.

Mr. Toby Selcer and Mr. Carl Fredericksen and the entire Texas A&M machine and electronics shops helped in the development and fabrication of the telltales and other needed devices.

Many Texas A&M University undergraduate students had a hand in setting up each of the load tests as well as collecting data during the tests (including many 24 hour, cold and wet late night readings). Particular thanks needs to go to students Juan Quiroz and Christine Evatt for their extra help.

Mr. Robert Gibbens, Masters Candidate at Texas A&M University and Engineer with Geotest Engineering was directly responsible for the load test and soil tests. It is his quality work in organizing, supervising and getting the job done which has made this project a success.

V.N. Vijayvergiya
Houston April 7, 1994

Introduction

Albert F. DiMillio,[1] Member, ASCE

Abstract

A brief description of the spread footing load test program is presented to orient the prediction symposium participants. Some background material is also presented to place the role of the prediction symposium in the proper perspective. A brief discussion of the foundation selection and design process used in highway engineering is also presented to inform the participants of the current state-of-the-practice. Previous and current Federal Highway Administration (FHWA) research on spread footings is described to demonstrate how the current load testing project fits into the overall research program.

Introduction

This prediction symposium was established to examine various design methodologies for spread footings on granular soils that are used by government agencies, consultants, and academia. The predictions have been compared to the measured results of load tests on actual spread footings which were conducted under a research contract between the FHWA and Geotest Engineering, Inc., of Houston, Texas. The tests were conducted at the National Geotechnical Experimental Site on the campus of Texas A&M University.

Five load tests on square footings ranging from 1x1 m to 3x3 m in size were conducted at a sandy site where numerous soil tests were performed. Vertical loads were applied at the center of the footings which rested on flat ground. Load and settlement data were recorded for each increment of load application. All of the predictors were supplied the same basic information in order to make a Type A prediction. No predicted or measured values were revealed before the symposium.

The load tests were a major part of a comprehensive study of spread footing behavior on granular soils. A computer-aided data base was developed to provide storage and retrieval capabilities for these

----

[1] Geotechnical Research Manager, Federal Highway Administration, 6300 Georgetown Pike, McLean, Virginia 22101-2296

load test data plus other data obtained from previous FHWA studies and those from the open literature market. The data base program also contains an analysis module for performing various prediction techniques commonly used by the geotechnical community.

The data base is mainly a research tool to assist in the evaluation and improvement of existing prediction techniques for bearing capacity and settlement of footings on sand; however, it should be a valuable tool for practioners as well. It is also quite possible that the data base will assist future researchers to develop a new method(s) that will more adequately model actual behavior of a spread footing on sand.

Background

The current contract study is the latest of a series of FHWA studies that began in the early eighties to improve the design procedures for spread footings used to support highway structures, and to demonstrate the dependability and cost-effectiveness of these foundation elements as a viable alternative to deep foundation systems.

The major decision to be made in foundation engineering, which is one of the oldest professions of mankind, has always been whether to use shallow or deep foundations to support a structure. Although spread footings are usually much less expensive than deep foundation systems, highway bridge engineers have normally opted to use deep rather than shallow systems because of a perception that they are more reliable and provide a higher safety factor. Uncertainties regarding the design and analysis of spread footing systems have also discouraged their use. As a result, the use of spread footings to support highway structures in the United States varies widely between States and is generally far below the optimum use level; hence the initiation of the FHWA research program on spread footings.

Some of the reasons piles are preferred by some designers over spread footings as a foundation for highway bridges include the following:

- The lack of well-documented performance evaluation data
- The lack of rational tolerable movement criteria for bridges
- Skepticism and uncertainties concerning the potential cost savings
- The questionable accuracy of settlement predictions for spread footings.
- Skepticism and uncertainties concerning the quality of fill or natural ground below the spread footings.

The current FHWA research program was designed to provide better information and data to allay the fears of bridge foundation designers concerning the above reasons.

Foundation Selection Process

The need for adequate bridge foundation support is well established, and a comprehensive and rational foundation selection process must be used to achieve adequate support.

Foundation economies occur mostly in the foundation selection process. If a designer can justify the use of spread footings for a particular situation, large cost reductions will usually occur. If a spread footing can support a bridge abutment or pier adequately, further cost savings will be minimal in the actual design of the footings. If a pile foundation is required, cost savings still are possible through a proper engineering analysis and selection of the type, size, and location of the piles.

The rational selection of a safe and economical foundation for bridge abutments and piers involves the relationships between subsurface conditions, structural loads, performance requirements, construction methods, and environmental effects. A systematic process of evaluating these factors is necessary to insure proper foundation selection.

Subsurface conditions must be investigated thoroughly to choose and design the most economical foundation. The lack of proper soil and site information is a major reason pile foundations are used so frequently for bridges because an engineer will tend to select piles if the design data are incomplete. If proper soils information is available, the primary concern in soil support for structures is volume stability, strength, and durability. However, adequate support is highly variable and is more case-specific than site-specific. In the proper state, virtually any soil type, except highly organic materials, may be adequate for foundation support. Conversely, any soil type, in its natural state, may be inadequate for foundation support. Adequate soil support depends on the previously mentioned relationships between factors such as structural loads and performance requirements.

Whether a given soil is adequate for foundation support and whether a given inadequate soil can or should be made adequate depend on the following factors that must be evaluated for each case:

- Soil type and present properties.
- Area and depth of soil treatment required.
- Local site conditions, including location of water table, presence of utilities, and obstructions of unknown location or size.
- Kind of structure to be supported.
- Level of improvement needed; for example, required bearing capacity, permissible settlements, and minimum relative density.
- Availability of foundation materials, skilled laborers, and equipment.
- Time available.

- Environmental factors, including waste disposal, ground water pollution, noise, and erosion.
- Cost relative to deep foundation alternatives.

A group of piles is not more stable than a spread footing except when the piles can transfer the load through a weaker soil layer(s) to a firm, unyielding ground layer; or, in the case of excessive depth to a firm layer, the piles are driven deep enough into the weak layer(s) to develop practical refusal from frictional resistance. The load must be balanced against available soil support and performance requirements established by the bridge engineers for the particular structure being designed.

If the estimated costs of various foundation alternatives (determined during the design stage to be feasible) are within 15 percent of each other, alternate foundation systems might be included in contract bid documents.

Spread Footing Design

The advent of reinforced concrete in the early 1900's and recent improvements in excavation technology have increased greatly the appeal of spread footings for bridge foundations. Modern soil mechanics and improved methods of site investigation and laboratory testing have improved the accuracy of settlement and bearing capacity predictions. Also, compaction control and improved grading procedures have minimized spread footings being founded on weak and/or compressible soils. Finally, special ground improvement techniques such as soil reinforcement, stone columns, and dynamic compaction have increased the attractiveness and applicability of spread footings.

Bearing capacity and settlement determine allowable foundation pressures. Usually, settlement is the controlling factor. In the design of spread footings, footing dimension proportions usually are based on 25 mm of settlement and a safety factor of 2.5 or 3 with respect to bearing capacity failure. The bearing capacity is calculated from the estimated shear strengths of the supporting soils using the standard Terzaghi bearing capacity factors. The magnitude and rate of settlement are computed from the compressibility properties of the foundation soils.

Some highway agencies have a policy not to use spread footings on cohesive soils because of the concern for bearing capacity failure and/or excessive settlement. This is a conservative approach because some cohesive soil deposits (especially overconsolidated clays) can support heavy bridge loads without distress resulting. Although bearing capacity on sands is not a problem, some highway agencies do not use spread footings on cohesionless soils because of the concern for excessive settlement.

This also is a conservative policy because a spread footing on sand usually will provide satisfactory support because consolidation of sands usually is minimal and occurs rapidly. Most of the settlement occurs before the sensitive superstructure elements are erected.

The use of spread footings on compacted fill also is infrequent. Although a properly compacted fill often is stronger and more stable than natural ground and easily able to support a spread footing, some designers use spread footings on in-situ soils and avoid their use on prepared fills. Large settlements from fills usually can be traced to older, nonuniform fills that may have been constructed of poor soils or uncompacted waste materials dumped on unprepared natural ground surfaces. However, the use of random, uncontrolled fill in modern highway construction, especially near bridges, is readily avoidable. The Connecticut and Washington State transportation departments frequently support bridge abutments on spread footings in compacted fill with little or no problems.

Preventing collapse of the spread footing is the main concern of the foundation designer; however, such failures are virtually nonexistent for highway bridges because of high safety factors. Failure also can be caused by excessive, long term settlement. Although collapse does not usually occur, excessive settlement can severely crack the abutments and piers, or it can overstress key superstructure elements such as girders and deckslabs. This kind of failure usually results from design error or improper construction rather than from an inherent feature of spread footings. All too often, though, a failure caused by collapse or excessive settlement unduly discourages highway engineers from further use of spread footings.

A foundation system must be functional as well as safe. There is a wide degree of engineering performance between an unyielding support system and one that fails. Persistent maintenance problems and failures of noncritical elements (such as parapet walls and joints) are expensive to correct and should be avoided if peculiar to certain systems, situations, or methodologies. To improve the design process, engineers should correlate functional distress (bumps, cracks, and misalinements) with system characteristics (abutment type, soil type, superstructure type, and amount and kind of movement) to determine where spread footings are not appropriate.

When a spread footing design is inappropriate or the soils in their natural state are not able to provide adequate support, it is common to use piles or drilled shafts. However, this can be costly because sometimes it is more economical to improve the soils rather than use piles or remove and replace the unsuitable soils. Ground improvement methods--such as vibroflotation, stone columns, soil reinforcement, and dynamic compaction--generally fall in the following categories:

- Compaction by vibration.
- Compaction by displacement and vibration.
- Grouting and injection.

- Precompression.
- Reinforcement.

Many ground improvement methods originating in Asia and Europe are gaining widespread acceptance in the United States. Some methods have been used in the highway industry after successful use in the building industry.
Many new ground improvement methods currently being examined by FHWA researchers are quicker, less expensive, more durable, and less disruptive to traffic operations than standard methods. For example, reinforced soil structures reduce total costs by using earth fill instead of concrete, by reducing site preparation and installation time, by increasing tolerance to differential settlements, and by providing low-maintenance structures.

FHWA Research

During the early stages of the FHWA research program, a study was made to develop rational criteria for tolerable movements of highway bridges. Along with the criteria, a basic design procedure was developed that permits a systems approach for designing the superstructure and the foundation system. This design procedure incorporates the tolerable movement guidelines that are based on strength and serviceability criteria which, in turn, are based on limiting longitudinal angular distortion, horizontal movements of abutments, and deck cracking.

To increase the number of documented case studies of spread footing performance, FHWA staff, in cooperation with the Washington State Department of Transportation, evaluated the performance of numerous highway bridge abutments supported by spread footings on compacted fill. During this review, the structural condition of 148 highway bridges throughout Washington State was visually inspected. The approach pavements and other bridge appurtenances also were inspected for damage or distress that could be attributed to the use of spread footings on compacted fill.

Based on this review and detailed investigations of the foundation movement of 28 selected bridges, it was concluded that spread footings can provide a satisfactory alternative to piles, especially when high embankments of good quality borrow materials are constructed over satisfactory foundation soils. None of the bridges investigated in Washington displayed any safety problems or serious functional distress; all bridges were in good condition.

In addition to the performance evaluation, cost-effectiveness analyses and tolerable movement correlation studies further substantiated the feasibility of using spread footings in lieu of expensive, deep foundation systems. Spread footings cost 50 to 65 percent less than pile foundations and foundation movement studies showed that the 28 bridges inspected easily tolerated differential settlements of 25 to 75 mm without serious distress.

FHWA subsequently initiated a 3-year comprehensive review of spread footing design and performance by studying, through the use of instrumentation, the performance of 10 bridges supported on spread footings as well as a review and evaluation of current methods to predict settlement of spread footings on cohesionless soils. The methods, generally either empirical or based on elastic theory, varied in complexity and the treatment of ground water level and footing size. This study identified the best methods available, and parallel research was conducted to refine and improve the selected methods.

During the last 10 years, a substantial amount of field performance information has been obtained and analyzed. Recent model tests on smaller footings from 1/3 m to 1 m performed at the FHWA Turner-Fairbanks Highway Research Center have added to the knowledge data base. These data and the Texas A&M test data will be used to refine the prediction methods selected in previous studies. Lessons learned from this prediction symposium and recent advances from the spread footing research program should reduce the skepticism, uncertainty, and concern for spread footing performance.

It is hoped that the description of the FHWA spread footing research program and the comments concerning the conservative approach taken by some highway agencies have provided sufficient background information to allow you to place the role of this prediction symposium in the proper perspective. We have some very interesting load test results to present for discussion. It is always beneficial to gather together a group of expert predictors and a knowledgeable audience to discuss and analyze load test results and design procedures. Type A prediction symposia like this one present a challenge and provide a forum where experts can showcase their talents and learn from each other. The proceedings of such an event can be very useful to other researchers and practioners too.

I would like to close with a few words about predictions. Predictions are a very important part of the decision making process. Our community needs to do more in the way of evaluating our prediction methods in order to improve the state-of-the-practice. Predictions are sometimes far off the mark, often due to faulty or insufficient information, but generally because our procedures are somewhat inaccurate. Large research quality data bases will help future researchers develop new and/or improved prediction methods.

DATA AND PREDICTION REQUEST FOR THE SPREAD FOOTING PREDICTION EVENT

SPONSORED BY FHWA
AT THE OCCASION OF THE ASCE
SPECIALTY CONFERENCE: SETTLEMENT '94
by

Robert Gibbens and Jean-Louis Briaud

SETTLEMENT '94
is the ASCE Specialty Conference on the topic of
Vertical and Horizontal Deformations of Foundations
and Embankments to be held at Texas A&M University
on June 16, 17 and 18, 1994

June 1993

ACKNOWLEDGMENTS

The following persons and companies are thanked for their contributions to this project:

This project is sponsored by the Federal Highway Administration in Washington. Mr. Al DiMillio and Mr. Michael Adams are the technical contacts for this project at FHWA. The project is being performed by Geotest Engineering in cooperation with Texas A&M University. Dr. Vijay Vijayvergiya is in charge of the project at Geotest Engineering, Inc., while Dr. Jean-Louis Briaud is in charge at Texas A&M University.

Dr. Silvano Marchetti and Mr. Maurizio Calabrese of the University of L'Aquila came all the way from Italy to perform the Dilatometer Tests free of charge and are considered to be cosponsors of the project.

Dr. Phillip Buchanan and his team at Buchanan/Soil Mechanics, Inc (B/SMI) provided, at a reduced rate, all the drilling rig services including those for Standard Penetration Tests, for Pressuremeter Tests, for Dilatometer Tests and for setting casings for Cross Hole Tests, Inclinometer Tests and Telltales; B/SMI is considered to be a cosponsor of the project.

Dr. George Goble of Goble Rausche Likins and Associates, Inc (GRL) came from Boulder to perform the SPT energy measurements and charged a very reduced rate; GRL is considered to be a cosponsor of the project.

Mr. Recep Yilmaz and his team at Fugro McClelland performed the PiezoCone Penetrometer Tests.

Dr. Allan Lutenegger and Dr. Don Degroot of the University of Massachusetts performed the Step Blade Test, the Borehole Shear Tests and provided the load cell for the axial thrust measurements during the Dilatometer Test. Mr. Michael Adams of FHWA performed the Borehole Shear Test along with Dr. Lutenegger and Dr. Degroot.

Dr. Ken Stokoe of the University of Texas provided the second Dilatometer apparatus used for the DMT sounding, performed independent of Mr. Calabrese's tests.

Dr. Derek Morris and Mr. Tony Yen of Texas A&M University performed the Resonant Column Tests and the Cross Hole Tests.

Mr. Phillipe Jeanjean of Texas A&M University performed the Triaxial Tests and the Pressuremeter Tests.

The second author wishes to thank the first author Mr. Robert Gibbens, Masters Candidate at Texas A&M University and Engineer with Geotest Engineering, for all his quality work in organizing, supervising and participating in the field and laboratory activities. His help was invaluable.

TABLE OF CONTENTS

1. INTRODUCTION

The purpose of this report is to give the results of the soil tests performed at one of the two National Geotechnical Experimentation Sites located on the Texas A&M University Riverside Campus for the spread footing prediction symposium. The work is sponsored by the Federal Highway Administration (Mr. Al DiMillio) and performed by Geotest Engineering (Dr. Vijayvergiya and Mr. Robert Gibbens) and by Texas A&M University (Dr. Jean-Louis Briaud).

The goal is to evaluate the profession's ability to predict the behavior, at small and large deflections, of footings on sand. To that end 5 square footings are being constructed at the sand site. The footings include two 3x3 m footings, one 2x2 m footing, one 1.5x1.5 m footing and one 1x1 m footing. All footings will be load tested to 0.15 m of settlement. The insitu soil tests which have been conducted at the site are: CPT, PMT, SPT w/Energy, DMT w/axial force, Step Blade, Borehole Shear and Cross-Hole Wave tests. The laboratory soil tests which have been conducted include Index Property Tests, Triaxial Tests and Resonant Column Tests. Those who conducted the tests are acknowledged in the section describing the test.

2. SPREAD FOOTING LOAD TEST SETUP

Five spread footings (Figure 1) will be load tested at the site: two 3 x 3 m footings, one 2 x 2 m footing, one 1.5 x 1.5 m footing and one 1 x 1 m footing. All footings will be founded at a depth of 0.76 m in the sand and will be 1.2 m thick.

The load will be applied vertically at the center of the footings. The loading sequence for all the footings will be as shown on Table 1.

The soil at the site will be instrumented with extensometers to measure vertical displacements and with inclinometer casings to measure horizontal displacements. A total of three extensometers per footing will be placed at depths of 0.5B, 1B and 2B. Inclinometer casings were installed before the footings to facilitate the cross-hole wave tests. The casings were placed in locations corresponding to the cross-hole tests, as shown on Figure 3.

Figure 2 shows a cross sectional view of the site. The site is bordered by a man-made embankment on one side and a cut slope on the other. The base of both slopes is over 6.5 m away from the closest edge of any footing.

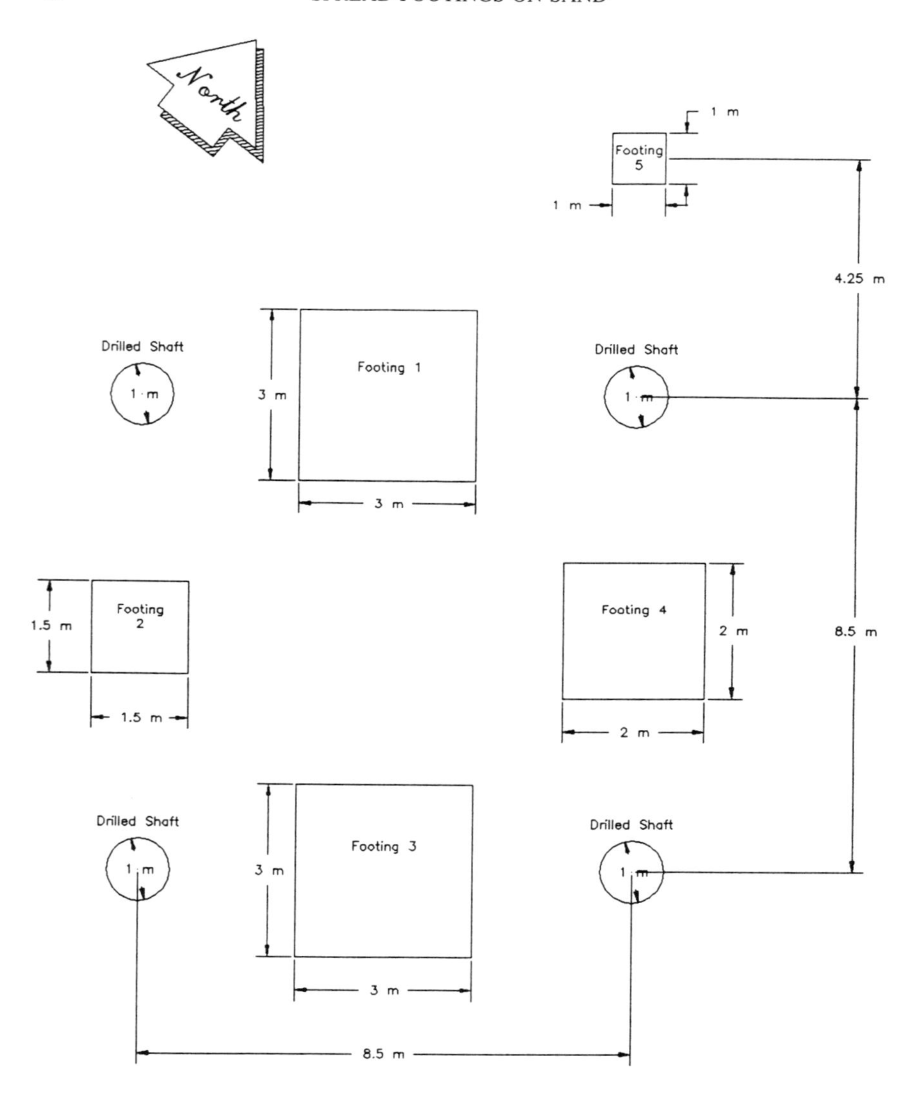

FIGURE 1
Footing/Pier Layout

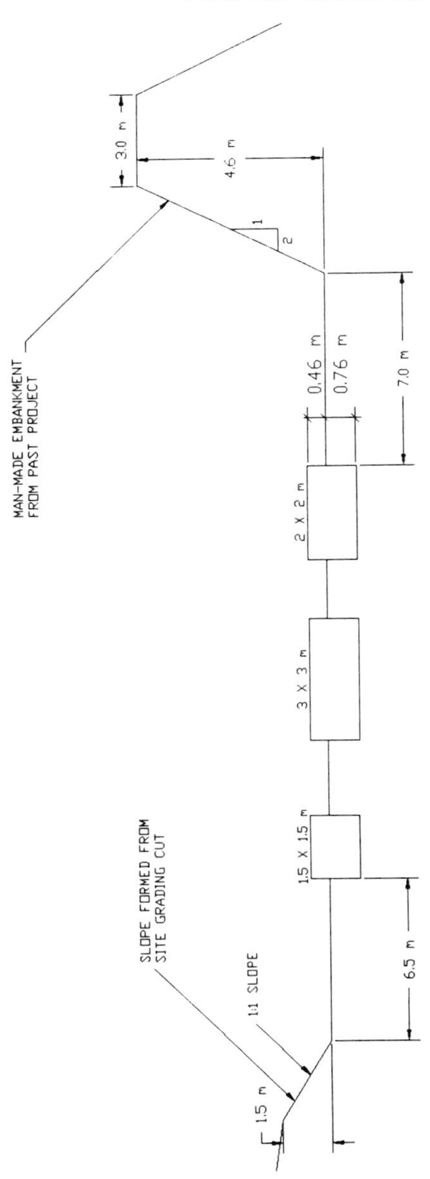

FIGURE 2
Site Cross Section

Day	Start Time	End Time	% of Estimated Load Capacity of the Footing	Reading Intervals	Time Units
1	08:00	08:30	0	1,3,6,10,20,30	minutes
	08:30	09:00	10	1,3,6,10,20,30	minutes
	09:00	09:30	20	1,3,6,10,20,30	minutes
	09:30	10:00	30	1,3,6,10,20,30	minutes
	10:00	10:15	30-0	continuous	minutes
	10:15	10:30	0-30	continuous	minutes
	10:30	11:00	40	1,3,6,10,20,30	minutes
	11:00		40	1,3,6,10,20,30	hours
Leave load for 24 hours					
2	11:00	11:30	50	1,3,6,10,20,30	minutes
	11:30	12:00	60	1,3,6,10,20,30	minutes
	12:00	12:30	70	1,3,6,10,20,30	minutes
	12:30	13:00	80	1,3,6,10,20,30	minutes
	13:00	13:30	90	1,3,6,10,20,30	minutes
	13:30	14:00	100	1,3,6,10,20,30	minutes
Footing may be loaded beyond estimated capacity					
	14:00	14:30	100-0	continuous	minutes
	14:30	15:00	0-100	continuous	minutes

TABLE 1
Loading Sequence For All Footings

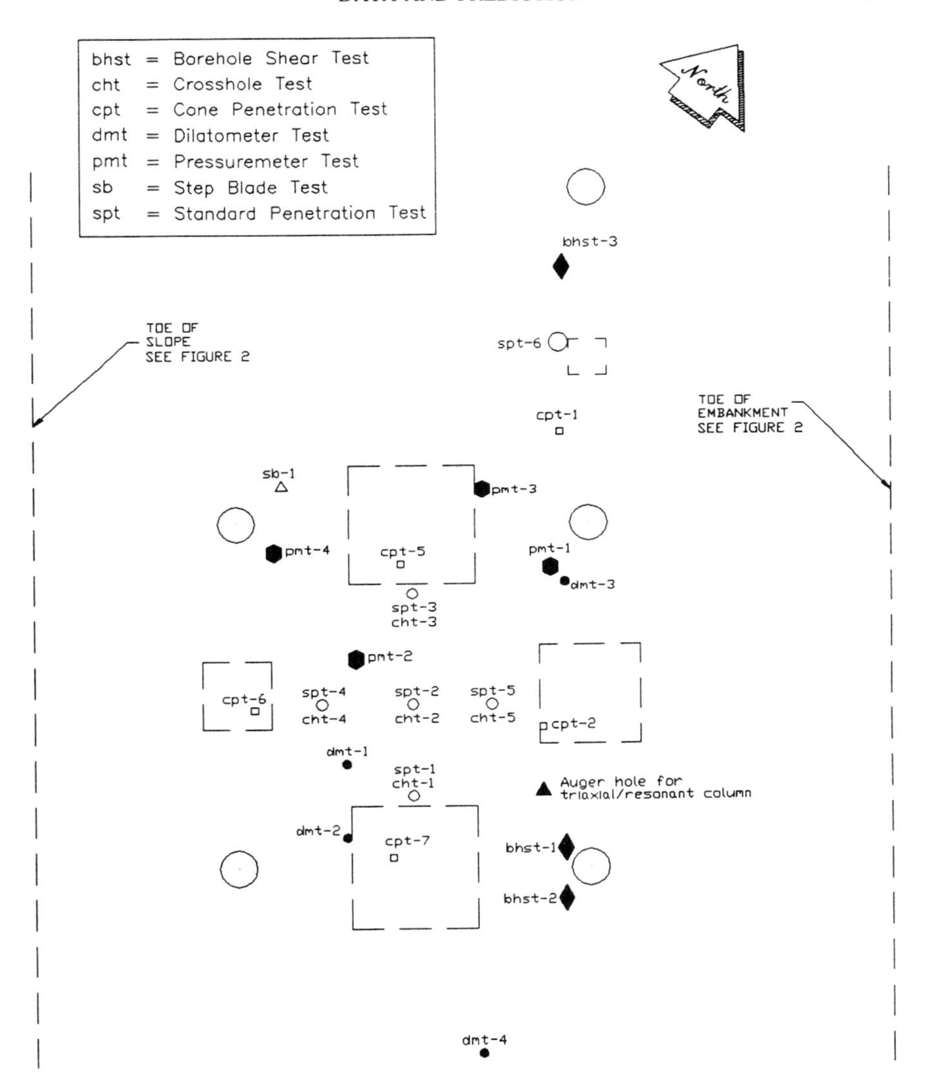

FIGURE 3
Field Testing Layout

3. SOIL DATA

3.1 General Soil Description:

The soil at the National Geotechnical Experimentation Site selected for this project is predominantly sand from 0 to 11 m. Below the sand layers is a clay layer which exists until a depth of at least 33 m. The water table was observed in an open well to be at a depth of 4.9 m. Figure 4 shows the general layering at the site. The sand is a medium dense silty fine sand and the clay is a very hard plastic clay. A detailed site investigation was undertaken to characterize the sand deposit (Figure 3 and Table 2).

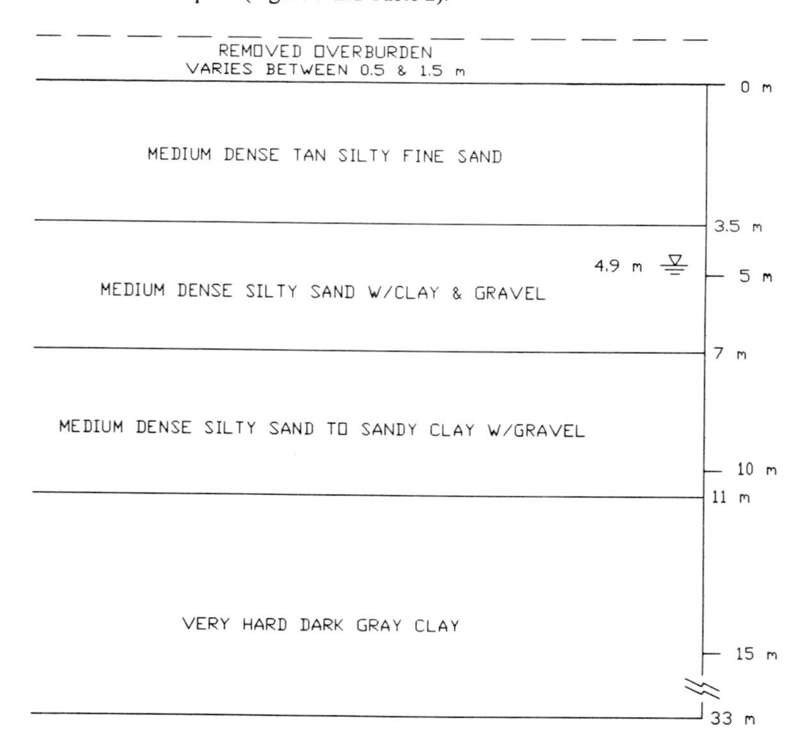

FIGURE 4
General Soil Layering

TEST	DATE	PEOPLE INVOLVED
Borehole Shear Tests	April 18 & 19, 1993	Mr. Mike Adams Dr. Allan Lutenegger Dr. Don Degroot Buchanan/Soil Mechanics
Cross-Hole Wave Tests	May 13, 1993	Dr. Derek Morris Mr. Tony Yen Mr. John Delphia
PiezoCone Penetration Test	January 27, 1993	Fugro McClelland
Dilatometer Tests	April 8 & 13, 1993	Mr. Maurizio Calabrese Mr. Robert Gibbens Buchanan/Soil Mechanics
Dilatometer Test w/Thrust Measurements	May 27, 1993	Mr. Robert Gibbens Buchanan/Soil Mechanics
Pressuremeter Tests	February 2, 1993 May 15, 1993	Dr. Jean-Louis Briaud Mr. Philippe Jeanjean Mr. Robert Gibbens Mr. Rajan Viswanathan Buchanan/Soil Mechanics
Step Blade Test	April 19, 1993	Dr. Allan Lutenegger Dr. Don Degroot Buchanan/Soil Mechanics
Standard Penetration Tests	April 5-21, 1993	Mr. Robert Gibbens Dr. George Goble Buchanan/Soil Mechanics
Water Content & Unit Weights	November 1992-May 1993	Mr. Philippe Jeanjean Mr. Robert Gibbens
Atterberg Limits	May 1993	Mr. Robert Gibbens
Relative Density	November 1992	Mr. Philippe Jeanjean
Triaxial Tests	April 1993	Mr. Philippe Jeanjean
Resonant Column Tests	May 1993	Dr. Derek Morris Mr. Tony Yen

TABLE 2
Test Dates & People Involved

3.2 Index Properties:

Visual classifications, sieve analysis and Atterberg limits were performed by Robert Gibbens of Geotest Engineering/Texas A&M University on SPT samples (SPT-2 through SPT-6 on Figure 3) and by Philippe Jeanjean of Texas A&M University on hand augered samples (boring labeled Auger hole on Figure 3). The "Gibbens" samples were obtained with a 51 mm split-spoon sampler as part of standard penetration tests. The boreholes were drilled by Buchanan/Soil Mechanics, Inc., using a Failing model 1500 truck mounted rotary drilling rig. The wet rotary method was used to advance or ream the borehole using a 121 mm wash bit and prepared drilling mud to keep the borehole from collapsing. The "Jeanjean" samples were obtained with a 76 mm hand auger in a dry hole.

Visual classifications were performed between 0 and 16.5 m at intervals described on the Boring Logs found in Figures 5 through 10. Atterberg limits were performed on a sample taken from boring SPT-2 at a depth of 16.4 m. The results indicate a liquid limit of 40, a plastic limit of 19 and a plasticity index of 21. Grain size analyses were performed on samples obtained from boring SPT-2 and from the hand augered hole . The tests were performed on samples ranging from 0 to 8 m at intervals described on the sieve analysis plots found on Figure 11. The locations of the borings can be found on Figure 3.

The test procedures followed for the grain size analysis and for the Atterberg limits conformed to ASTM procedures ASTM C136 and ASTM D4318, respectively.

BORING LOG

PROJECT: PERFORMANCE OF FOOTINGS ON SAND

BORING NO: SPT-1
LOCATION: SAND SITE

CLIENT: FEDERAL HIGHWAY ADMINISTRATION
DATE: 4/5/93
DRILLER: GUSTAVUS

PROJECT NO: 14G604
SOIL TECHNICIAN: GIBBENS

BORING TYPE: 121 mm BIT
GROUND ELEV:

▮-Shelby Tube Sample X-Penetration Sample J-Jar /-No Recovery
\\-Disturbed sample from cuttings ∇-Water encountered while drilling
-Open-hole water level B/F-Blows per foot, ASTM D 1586 Penetration Test
Pen.(tsf)-Field estimate of compressive strength

Depth in Meters	Blows Per 150mm/150mm/150mm	DESCRIPTION OF STRATUM
	4/5/6	NO SAMPLING
	7/9/14	NO SAMPLING
1.5	13/18/12	NO SAMPLING
	6/10/11	NO SAMPLING
	8/10/13	NO SAMPLING
3.0		
4.5	11/14/14	NO SAMPLING
6.0	9/15/19	NO SAMPLING
7.5	8/8/9	NO SAMPLING
9.0	4/5/8	NO SAMPLING
10.5	10/20/34	NO SAMPLING
12.0	19/29/47	NO SAMPLING
13.5	14/19/21	NO SAMPLING
15.0	15/22/31	NO SAMPLING
		Bottom @ 15.2 m

FIGURE 5
SPT-1

BORING LOG

PROJECT: PERFORMANCE OF FOOTINGS ON SAND
BORING NO: SPT-2
LOCATION: SAND SITE
CLIENT: FEDERAL HIGHWAY ADMINISTRATION
DATE: 4/6/93
PROJECT NO: 14G604
BORING TYPE: 121 mm Bit
DRILLER: GUSTAVUS
SOIL TECHNICIAN: GIBBENS
GROUND ELEV:

▌-Shelby Tube Sample **X**-Penetration Sample **J**-Jar **/**-No Recovery
\-Disturbed sample from cuttings **∇**-Water encountered while drilling
-Open-hole water level **B/F**-Blows per foot, ASTM D 1586 Penetration Test
Pen.(tsf)-Field estimate of compressive strength

Depth in Meters	Blows Per 150mm/150mm/150mm	DESCRIPTION OF STRATUM
	4/5/7	Tan Silty Fine Sand
	7/10/13	Tan Silty Fine Sand
1.5	8/8/10	Tan Silty Fine Sand
	6/9/9	Tan Silty Fine Sand
	4/8/8	Tan Silty Fine Sand
3.0		
	10/10/9	Tan Sand
4.5		
	7/9/8	Tan Sand w/Gravel
	7/10/11	Tan Sandy Clay
6.0		
	5/6/8	Tan Sandy Clay
7.5		
	6/8/13	Tan Silty Fine Sand
9.0		
	11/24/39	Dark Gray Clay
10.5		
	11/16/24	Dark Gray Clay
12.0		
	14/17/22	Dark Gray Clay
13.5		
	18/25/32	Dark Gray Clay
15.0		
		Bottom @ 15.2 m

FIGURE 6
SPT-2

BORING LOG

PROJECT: PERFORMANCE OF FOOTINGS ON SAND
BORING NO: SPT-3
LOCATION: SAND SITE

CLIENT: FEDERAL HIGHWAY ADMINISTRATION
DATE: 4/21/93
DRILLER: GUSTAVUS
PROJECT NO: 14G604
SOIL TECHNICIAN: GIBBENS
BORING TYPE: 121 mm BIT
GROUND ELEV:

█-Shelby Tube Sample X-Penetration Sample J-Jar /-No Recovery
\-Disturbed sample from cuttings ∇-Water encountered while drilling
-Open-hole water level **B/F**-Blows per foot, ASTM D 1586 Penetration Test
Pen.(tsf)-Field estimate of compressive strength

Depth in Meters	Blows Per 150mm/150mm/150mm	DESCRIPTION OF STRATUM
	5/5/8	Tan Silty Fine Sand
	4/8/10	Tan Silty Fine Sand
1.5	6/11/14	Tan Silty Fine Sand
	6/9/8	Tan Silty Fine Sand
	6/8/10	Tan Silty Fine Sand
3.0		
	4/9/10	Tan Sand
4.5		
	6/12/14	Tan Sand w/Gravel
	7/11/11	Tan Sandy Clay w/Gravel
6.0		
	6/9/10	Tan Sandy Clay w/Gravel
7.5		
	3/5/5	102 mm of Tan Silty Fine Sand becomes Tan Clay w/Gravel
9.0		
	11/20/24	26 mm of Clayey Gravel becomes Dark Gray Clay
10.5		
	43/48/51	Dark Gray Clay
12.0		
	14/18/28	Dark Gray Clay
13.5		
	13/15/21	Dark Gray Clay
15.0		Bottom @ 15.2 m

FIGURE 7
SPT-3

BORING LOG

PROJECT: PERFORMANCE OF FOOTINGS ON SAND
BORING NO: SPT-4
LOCATION: SAND SITE

CLIENT: FEDERAL HIGHWAY ADMINISTRATION
DATE: 4/16/93
PROJECT NO: 14G604
BORING TYPE: 121 mm BIT
DRILLER: GUSTAVUS
SOIL TECHNICIAN: GIBBENS
GROUND ELEV:

▌-Shelby Tube Sample X-Penetration Sample J-Jar /-No Recovery
\-Disturbed sample from cuttings ∇-Water encountered while drilling
-Open-hole water level B/F-Blows per foot, ASTM D 1586 Penetration Test
Pen.(tsf)-Field estimate of compressive strength

Depth in Meters	Blows Per 150mm/150mm/150mm	DESCRIPTION OF STRATUM
	3/4/7	Tan Silty Fine Sand
	5/8/7	Tan Silty Fine Sand
1.5	5/8/10	Tan Silty Fine Sand
	5/8/9	Tan Silty Fine Sand
	5/8/8	Tan Silty Fine Sand
3.0		
	6/8/7	Tan Sand
4.5		
	8/7/7	Tan Silty Fine Sand
	6/8/9	Tan Silty Fine Sand
6.0		
	8/8/8	Gravel
7.5		
	5/9/11	Tan Sandy Clay
9.0		
	11/23/27	76 mm of Tan Sandy Clay becomes Dark Gray Clay
10.5		
	21/31/39	Dark Gray Clay w/Gravel
12.0		
	14/21/32	Dark Gray Clay
13.5		
	17/23/37	Dark Gray Clay
15.0		
		Bottom @ 15.2 m

FIGURE 8
SPT-4

BORING LOG

PROJECT: PERFORMANCE OF FOOTINGS ON SAND
BORING NO: SPT-5
LOCATION: SAND SITE

CLIENT: FEDERAL HIGHWAY ADMINISTRATION
DATE: 4/13/93
DRILLER: GUSTAVUS
PROJECT NO: 14G604
SOIL TECHNICIAN: GIBBENS
BORING TYPE: 121 mm BIT
GROUND ELEV:

Depth in Meters	Blows Per 150mm/150mm/150mm	▌-Shelby Tube Sample X-Penetration Sample J-Jar /-No Recovery \\-Disturbed sample from cuttings ∇-Water encountered while drilling -Open-hole water level **B/F**-Blows per foot, ASTM D 1586 Penetration Test **Pen.(tsf)**-Field estimate of compressive strength
		DESCRIPTION OF STRATUM
	4/5/6	Tan Silty Fine Sand
	5/7/8	Tan Silty Fine Sand
1.5	6/10/10	Tan Silty Fine Sand
	4/8/11	Tan Silty Fine Sand
	4/7/9	Tan Silty Fine Sand
3.0		
	4/7/9	Tan Silty Fine Sand
4.5		
	8/12/15	Tan Silty Sand w/Gravel Pockets
	6/7/11	Tan Sandy Clay w/Gravel
6.0		
	4/9/7	Tan Sand
7.5		
	4/12/13	102 mm of Tan Sandy Clay w/Gravel becomes Gray Clayey Sand
9.0		
	12/33/31	51 mm of Fine Sand w/Gravel becomes Dark Gray Clay
10.5		
	11/17/21	Dark Gray Clay
12.0		
	10/14/22	Dark Gray Clay
13.5		
	11/12/27	Dark Gray Clay
15.0		Bottom @ 15.2 m

FIGURE 9
SPT-5

BORING LOG

PROJECT: PERFORMANCE OF FOOTINGS ON SAND
BORING NO: SPT-6
LOCATION: SAND SITE
CLIENT: FEDERAL HIGHWAY ADMINISTRATION
DATE: 4/21/93
PROJECT NO: 14G604
BORING TYPE: 121 mm BIT
DRILLER: GUSTAVUS
SOIL TECHNICIAN: GIBBENS
GROUND ELEV:

▌-Shelby Tube Sample X-Penetration Sample J-Jar /-No Recovery
\-Disturbed sample from cuttings ∇-Water encountered while drilling
-Open-hole water level B/F-Blows per foot, ASTM D 1586 Penetration Test
Pen.(tsf)-Field estimate of compressive strength

Depth in Meters	Blows Per 150mm/150mm/150mm	DESCRIPTION OF STRATUM
	6/6/7	Tan Silty Fine Sand
	6/8/11	Tan Silty Fine Sand
1.5	5/9/9	Tan Silty Fine Sand
	4/7/6	Tan Silty Fine Sand
	4/7/7	Tan Sand
3.0		
	11/11/15	Tan Sand
4.5		
	5/9/14	Tan Sand w/Gravel
	5/9/11	Gravel w/Sand
6.0		
	7/13/15	Tan Sand w/Traces of Gravel
7.5		
	3/4/4	No Recovery
9.0		
	9/20/29	76 mm of Clayey Gravel becomes Dark Gray Clay
10.5		
	12/20/31	Dark Gray Clay
12.0		
	16/18/35	Dark Gray Clay
13.5		
	17/21/33	Dark Gray Clay
15.0		Bottom @ 15.2 m

FIGURE 10
SPT-6

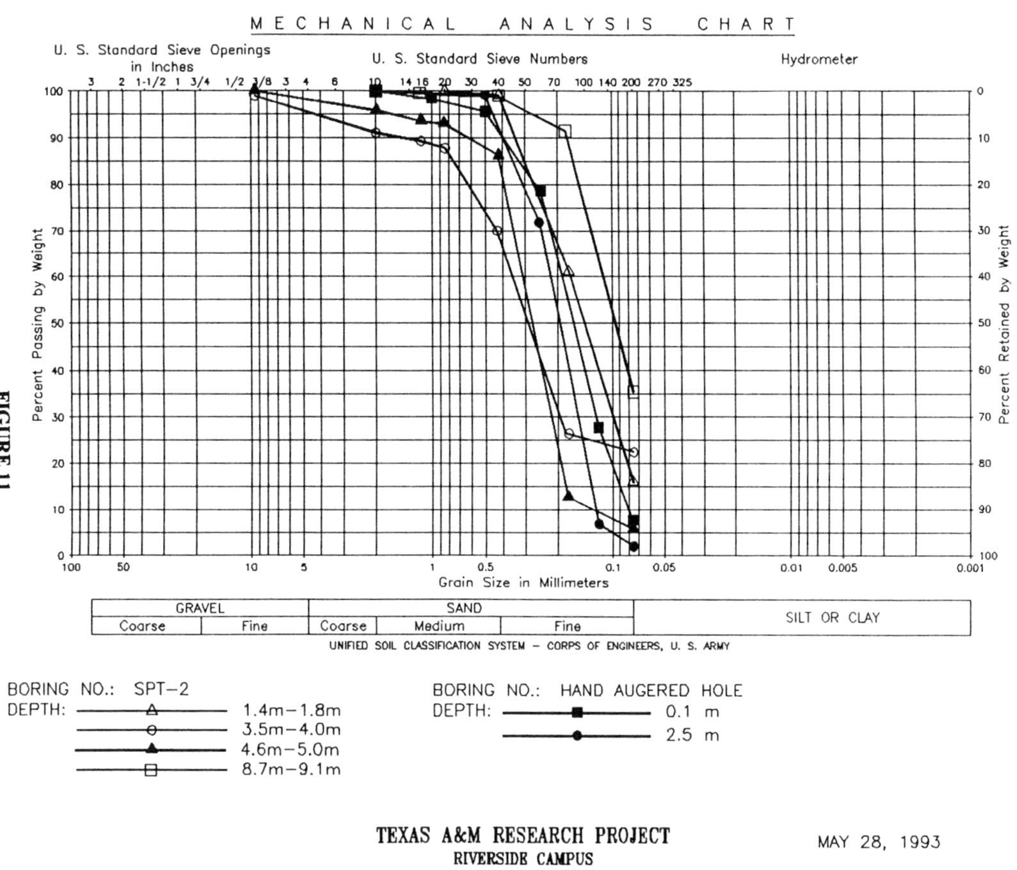

FIGURE 11
Grain Size Analysis

3.3 Water Content, Relative Density, Unit Weight

Water content, relative density and unit weight tests were performed by Philippe Jeanjean (G_s, w, e_{min}, e_{max}) of Texas A&M University and Robert Gibbens (w) of Geotest Engineering Inc./Texas A&M University. Table 3 gives the results of the tests run on the "Jeanjean" samples and Tables 4 through 6 give a summary of the moisture content results obtained from the "Gibbens" samples.

There is some variation with the moisture content; the numbers obtained in November 1992 after the dry summer in the hand augered hole are much lower than those obtained in April 1993 after the wet spring in the wet rotary borehole. The load test will be carried out during the late summer months. The higher water content obtained in April 1993 may be due to the rain or to the drilling mud or both.

The minimum and maximum void ratios were determined in accordance with ASTM D4254 and D4253 respectively and are given in Table 3. The insitu relative density was estimated from SPT blow counts and CPT values as being 55%. This, combined with the water content of 5%, leads to the natural unit weights shown on Table 3.

Property	Sand 0.6 m	Sand 3.0 m
Specific Gravity	2.64	2.66
Minimum Void Ratio	0.65	0.62
Maximum Void Ratio	0.94	0.91
Maximum Dry Unit Weight (kN/m^3)	15.70	16.10
Minimum Dry Unit Weight (kN/m^3)	13.35	13.66
Liquid Limit	N/P	N/P
Plastic Limit	-	-
USCS Classification	SP	SP-SM
Natural Void Ratio ?	0.78	0.75
Dry Unit Weight ? (kN/m^3)	14.55	14.91
Natural Moisture Content (%)	5.0	5.0
Natural Unit Weight ? (kN/m^3)	15.28	15.65

TABLE 3
Dry Hand Augered Samples: Test Results

BORING NO.	DEPTH IN METERS	MOISTURE CONTENT (%)
SPT-2	0-0.46	14.3
	0.76-1.2	13.0
	1.4-1.8	17.2
	2.0-2.4	19.2
	2.6-3.0	21.8
	3.5-4.0	16.3
	4.7-5.2	16.6
	5.6-6.1	27.7
	7.2-7.6	29.2
	8.7-9.1	27.4
	10.2-10.7	27.0
	11.7-12.2	27.5
	13.3-13.7	27.2
	14.8-15.2	22.1

BORING NO.	DEPTH IN METERS	MOISTURE CONTENT (%)
SPT-3	0-0.46	11.9
	0.76-1.2	12.1
	1.4-1.8	15.5
	2.0-2.4	18.3
	2.6-3.0	21.5
	3.5-4.0	21.1
	4.7-5.2	19.2
	5.6-6.1	25.4
	7.2-7.6	29.3
	8.7-9.1	23.0
	10.2-10.7	30.9
	11.7-12.2	34.0
	13.3-13.7	21.9
	14.8-15.2	25.8

TABLE 4
Wet Rotary Samples for SPT-2 & 3: Water Content Results

BORING NO.	DEPTH IN METERS	MOISTURE CONTENT (%)
SPT-4	0-0.46	16.8
	0.76-1.2	11.6
	1.4-1.8	11.8
	2.0-2.4	14.1
	2.6-3.0	19.0
	3.5-4.0	20.9
	4.7-5.2	20.7
	5.6-6.1	
	7.2-7.6	33.5
	8.7-9.1	20.9
	10.2-10.7	32.6
	11.7-12.2	36.7
	13.3-13.7	28.5
	14.8-15.2	21.7

BORING NO.	DEPTH IN METERS	MOISTURE CONTENT (%)
SPT-5	0-0.46	14.4
	0.76-1.2	13.0
	1.4-1.8	14.0
	2.0-2.4	20.5
	2.6-3.0	17.6
	3.5-4.0	16.6
	4.7-5.2	16.6
	5.6-6.1	19.4
	7.2-7.6	29.7
	8.7-9.1	29.6
	10.2-10.7	26.7
	11.7-12.2	29.3
	13.3-13.7	32.1
	14.8-15.2	25.1

TABLE 5
Wet Rotary Samples for SPT-4 & 5: Water Content Results

BORING NO.	DEPTH IN METERS	MOISTURE CONTENT (%)
SPT-6	0-0.46	15.9
	0.76-1.2	12.4
	1.4-1.8	11.1
	2.0-2.4	20.5
	2.6-3.0	18.9
	3.5-4.0	18.3
	4.7-5.2	18.7
	5.6-6.1	12.2
	7.2-7.6	31.2
	8.7-9.1	-
	10.2-10.7	28.6
	11.7-12.2	29.0
	13.3-13.7	30.7
	14.8-15.2	23.4

TABLE 6
Wet Rotary Samples for SPT-6: Water Content Results

3.4 Triaxial Tests on Reconstructed Samples:

Triaxial Tests were performed by Philippe Jeanjean of Texas A&M University on reconstructed samples obtained from the hand augered hole on Figure 3. The samples were reconstructed by matching the insitu water content and the estimated insitu relative density. The samples were built up in small layers using light tamping. The samples were 38 mm in diameter and 83 mm in height.

The tests were Consolidated/Drained triaxial tests, conforming to ASTM D3999 and were run at three different confining pressures: 34.5 kPa, 138 kPa and 345 kPa. The relative density of the sample was 55%. This was the insitu relative density estimated from SPT and CPT results. The water content was 5%.

The stress strain curves and the volume change curves obtained are presented in Figures 12 through 15. Note that a negative volumetric strain means a compression of the sample during shear. The values of the deviatoric stress at failure for each test are listed in Table 7. The corresponding Mohr's circles are shown on Figure 16. The calculated values of ϕ are:

at 0.6 m depth: $\phi = 34.2$ degrees

at 3.0 m depth: $\phi = 36.4$ degrees

Sample Depth (m)	Confining Pressure (kPa)	Deviatoric Stress at Failure $(\sigma_1-\sigma_3)_f$ (kPa)
0.6	34	83
0.6	138	503
0.6	345	885
3.0	34	109
3.0	138	420
3.0	345	103

TABLE 7
Triaxial Test Results

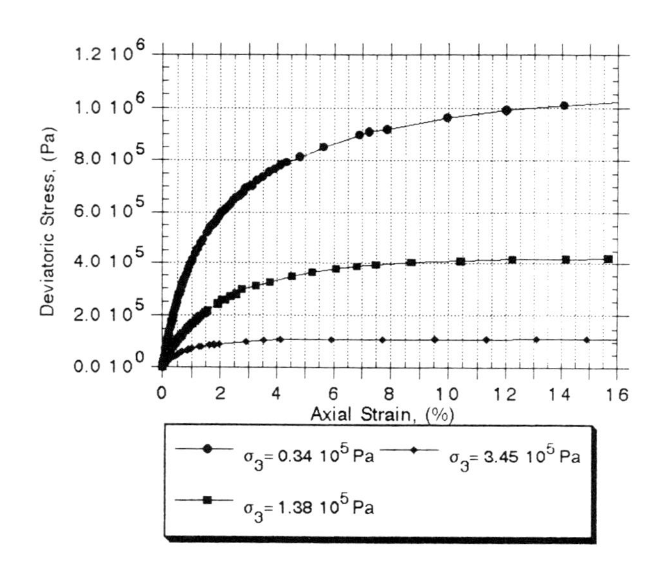

FIGURE 12
Stress Strain Curve for 0.6 m sample

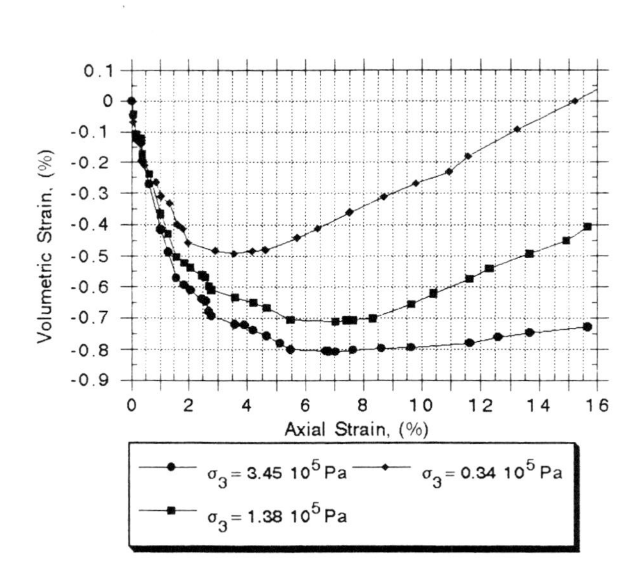

FIGURE 13
Volume Change Curve for 0.6 m sample

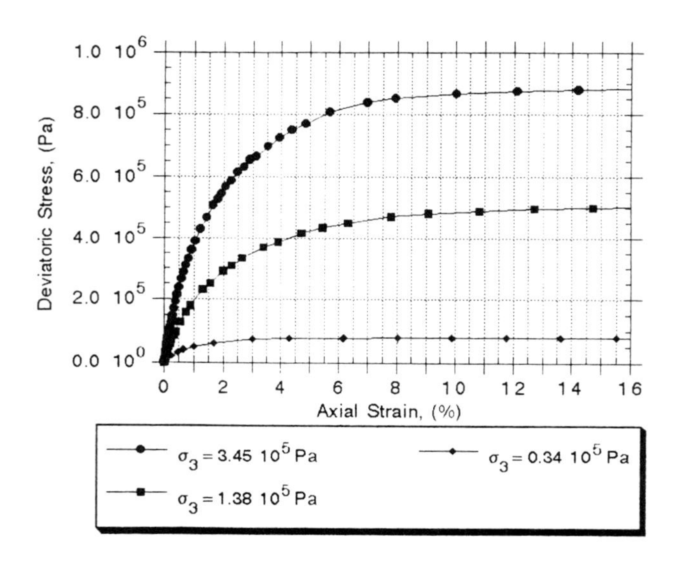

FIGURE 14
Stress Strain Curve for 3.0 m Sample

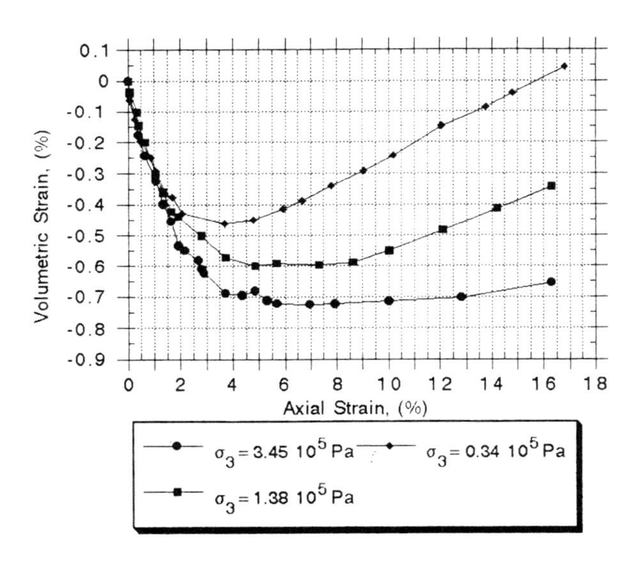

FIGURE 15
Volume Change Curve for 3.0 m Sample

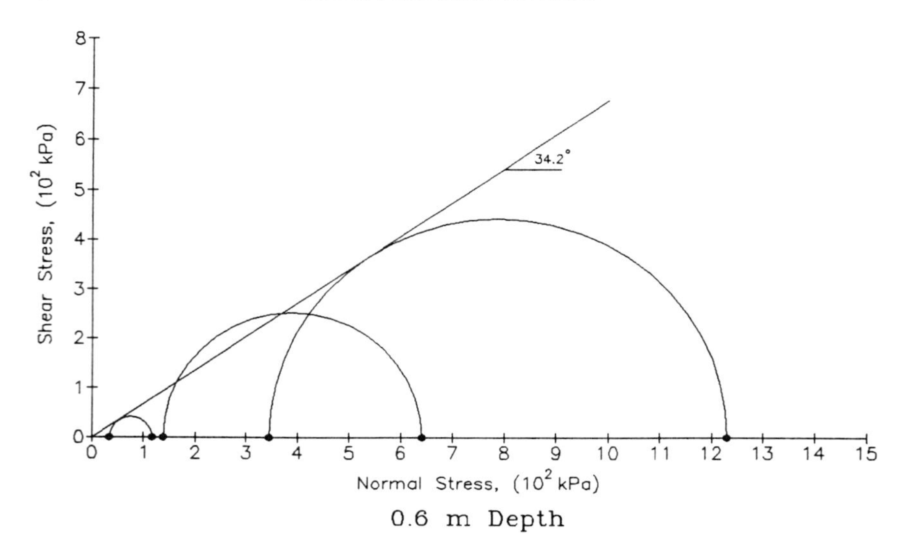

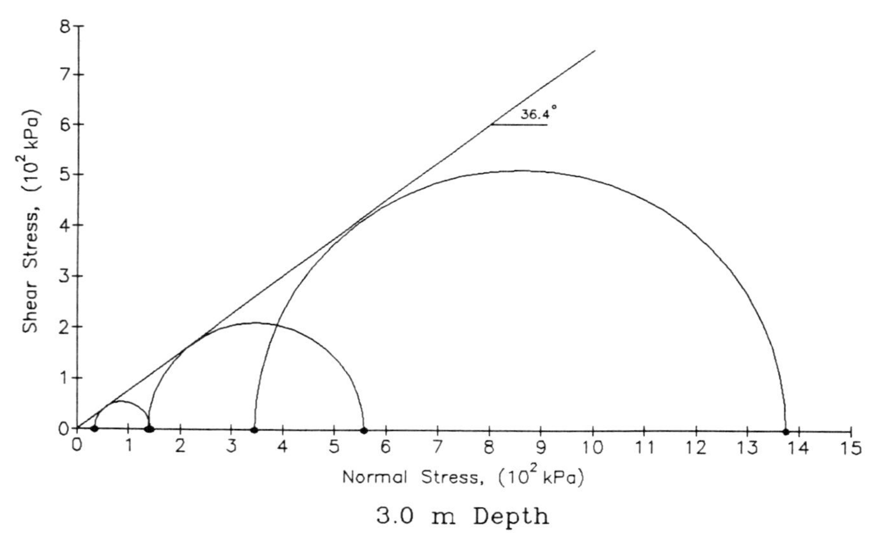

FIGURE 16
Mohr's Circles from Triaxial Tests

3.5 Resonant Column Tests on Reconstructed Samples:

Four Resonant Column Tests were performed by Derek Morris and Tony Yen at Texas A&M University. Three samples were reconstructed using sand from the samples taken in a dry hole with a hand auger at depths of approximately 0 m, 1.6 m and 3.3 m at the location shown on Figure 3. One sample was reconstructed using sand from SPT-2 at a depth of 6 m (Figure 3).

Each tested sample was reconstructed by compaction in the lab to approximately the same moisture content (5%) and relative density (55%) as originally found in the hand augered hole. The relative density of the sand at the site (55%) was estimated from SPT and CPT results. Each of the four samples was tested under six different confining pressures: 20, 50, 100, 150, 200 and 300 kPa. Due to the presence of approximately 10-20% by weight of particles passing the #200 sieve in the samples, a 24 hour consolidation period was used before each test. The first three samples had a diameter of 70 mm, the fourth sample had a diameter of 35 mm. The sample length was two diameters.

Tests were conducted in accordance with ASTM D4015 by measuring torsional resonance in an SBEL Stokoe resonant column device. This allowed the dynamic shear modulus to be computed for each sample, not only for different confining pressures, but also for at least one order of magnitude variation of shear strain. The compression modulus was not specifically measured, as it is very sensitive to the degree of saturation of the soil (and this particular equipment is not designed to measure it). However, it could be calculated if desired, assuming typical values of Poissons' ratio - typically 0.4 above the water table and 0.49 below the water table.

The results of the resonant column tests are shown as Shear Modulus G versus Shear Strain γ on figures 17 through 20. Tabulated values of G_{max}, the shear modulus measured at the smallest shear strain for a given confinement level, and soil property data can be found in Table 8.

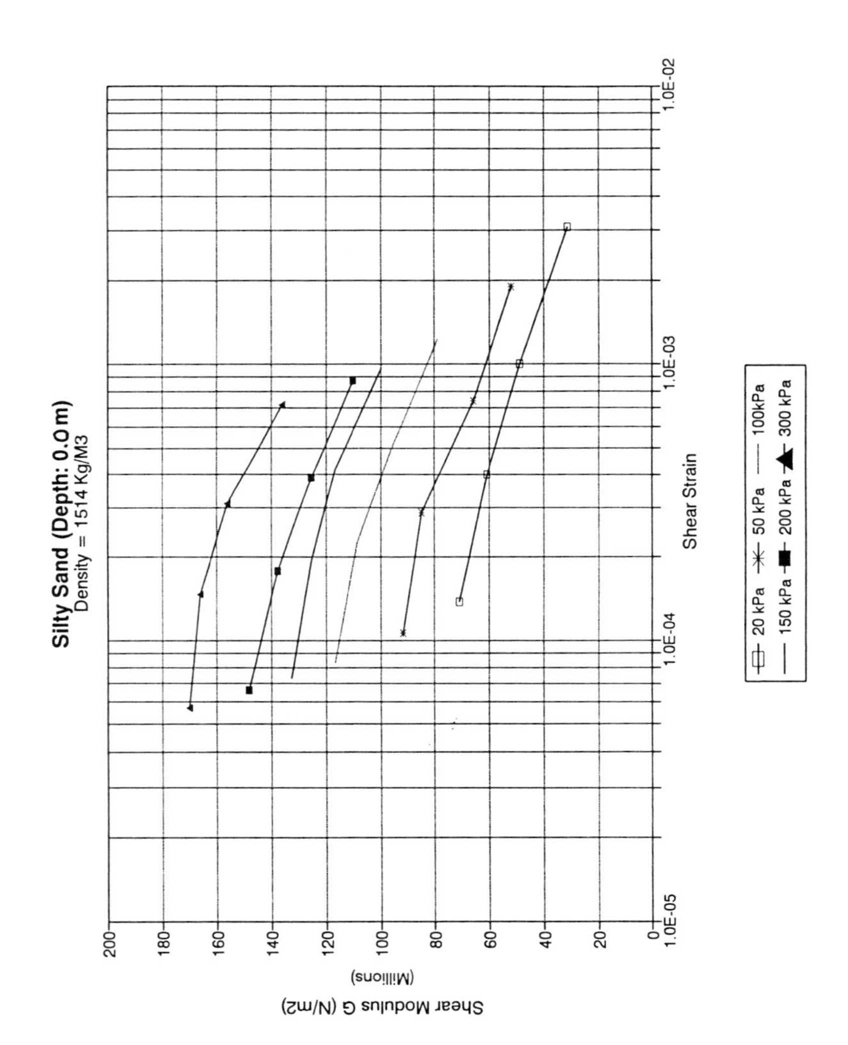

FIGURE 17
Resonant Column Test Results

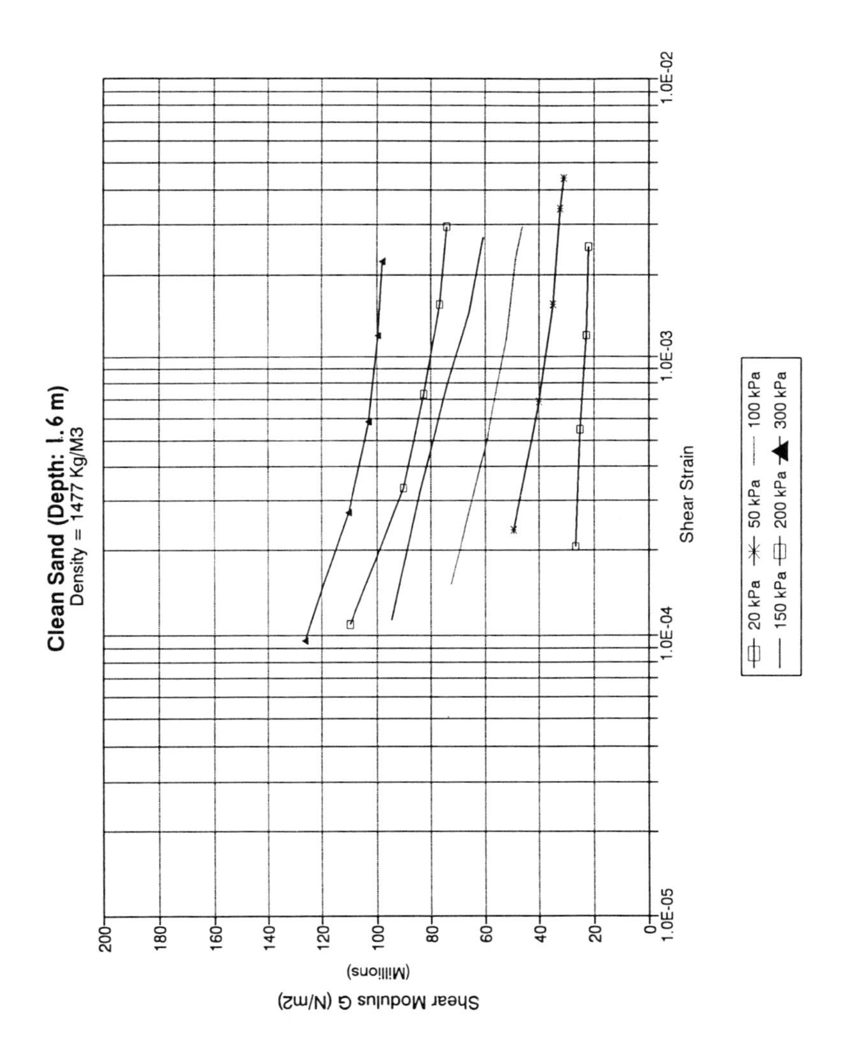

FIGURE 18
Resonant Column Test Results

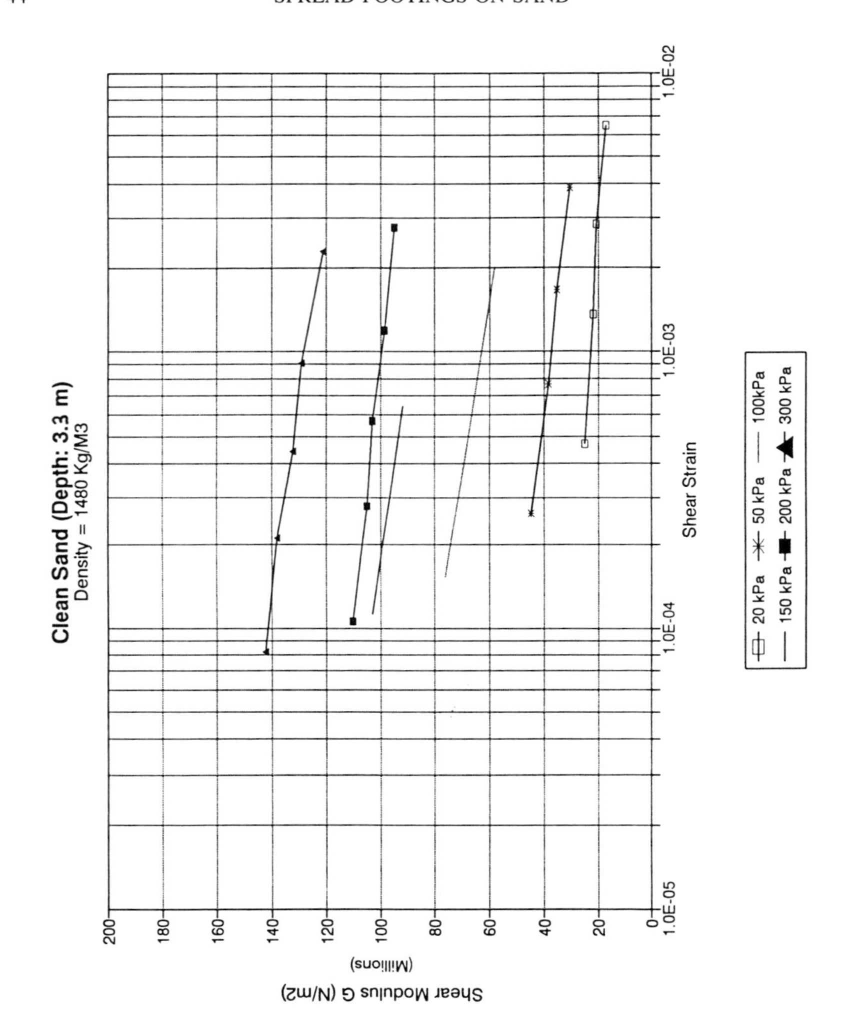

FIGURE 19
Resonant Column Test Results

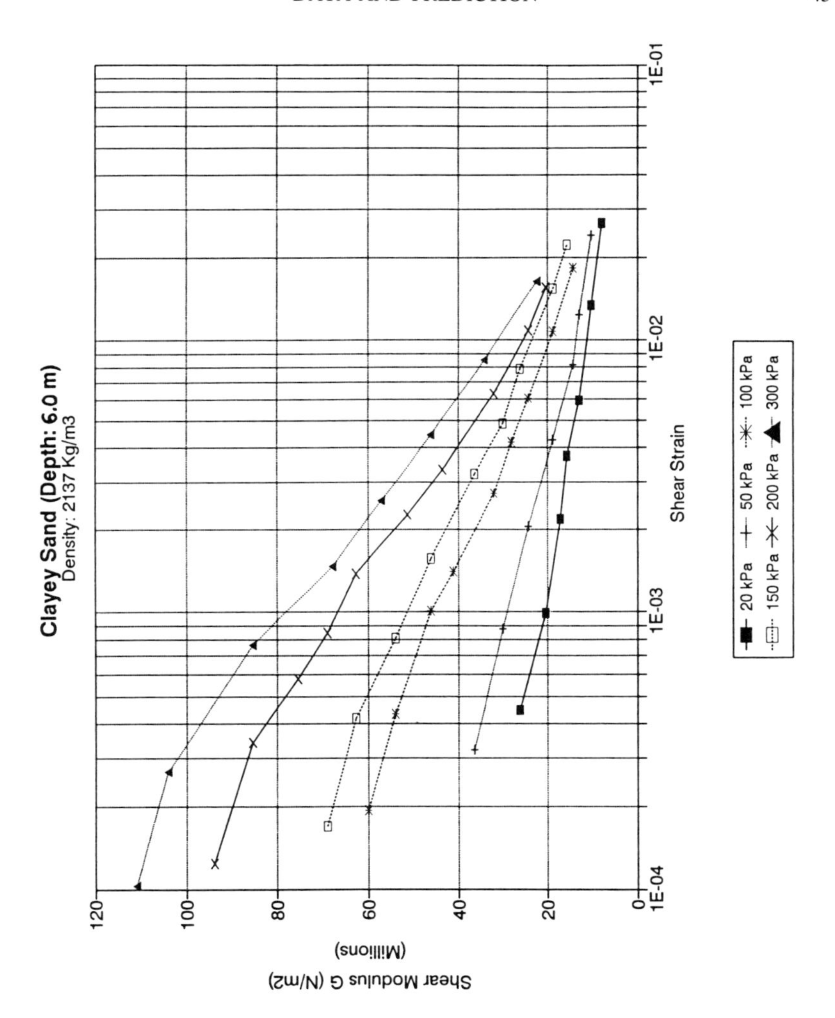

FIGURE 20
Resonant Column Test Results

SAMPLE 1	Silty Sand Depth 0 m. Density 1514 kg/m^3, water content adjusted to 5%					
Confining Pressure (kPa)	20	50	100	150	200	300
G_{max} (MPa)	71	92	117	133	148	170

SAMPLE 2	Clean Sand Depth 1.6 m. Density 1477 kg/m^3, water content adjusted to 5%					
Confining Pressure (kPa)	20	50	100	150	200	300
G_{max} (MPa)	26	49	72	95	110	126

SAMPLE 3	Clean Sand Depth 3.3 m. Density 1480 kg/m^3, water content adjusted to 5%					
Confining Pressure (kPa)	20	50	100	150	200	300
G_{max} (MPa)	25	44	76	103	111	142

SAMPLE 4	Clayey Sand Depth 6.0 m. Density 2137 kg/m^3, water content adjusted to 15%					
Confining Pressure (kPa)	20	50	100	150	200	300
G_{max} (MPa)	71	92	117	133	148	170

TABLE 8
Tabulated G_{max} Results

3.6 Standard Penetration Tests with Energy Measurements:

Standard Penetration Tests were performed by Buchanan/Soil Mechanics, Inc. and Geotest Engineering (Robert Gibbens). The tests were performed using a Failing Model 1500 truck mounted rotary drilling rig using a 623 N safety hammer, dropped a distance of 760 mm all in accordance with ASTM D1586. The drill rods used for this test were standard N rods in 3 m lengths with a 1.5 m length used as needed. A 121 mm wash bit was used to advance or ream the borehole prior to testing together with prepared drilling mud to keep the borehole from collapsing during testing. Six Standard Penetration test borings (Figure 3) were performed at the site, each to a final depth of 15.2 m. Blow counts over 450 mm of penetration were recorded in three 150 mm increments. The blow count N was taken as the sum of the blows for the last two 150 mm of penetration. These blow counts N are presented as a function of depth on Figures 21 and 22. The exact numbers are listed on the boring logs (Figures 5 through 10).

SPT w/energy measurements

One of the six SPT tests (SPT-2) was performed with energy measurements, while parts of another (SPT-1) was used for initial calibration and problem shooting. The energy measurements were made by George Goble of Goble Rausche Likins and Associates, Inc. (GRL) and the following text is a slightly modified excerpt from Goble's detailed report. The hammer velocity, the acceleration of the top of the drill string and the strain at the top of the drill string were measured during the SPT test and were used to determine the results presented on Table 9.

The dynamic measurements were obtained with a pair of piezoresistive accelerometers and with foil resistive strain gauges mounted at the midpoint of a 0.6 m long section (sub) of SPT drive rod. The sub was inserted into the drive string above the ground surface under the anvil. Analog signals from the gauges were conditioned, digitized, and processed with a Model PAK, Pile Driving Analyzer^{TM} (PDA).

The HPA radar was used to measure the SPT ram velocity. The HPA displays the ram speed as a function of time on a strip chart. A review of the strip chart provides ram impact velocity from which one can calculate the kinetic energy in the ram just prior to impact. The impact velocities were read manually from the strip chart in the laboratory.

The average energy delivered per blow during the boring SPT-2 was 0.25 kN-m. The maximum possible energy is 0.47 kN-m. Therefore the blow counts reported are measured with an energy efficiency averaging 53%.

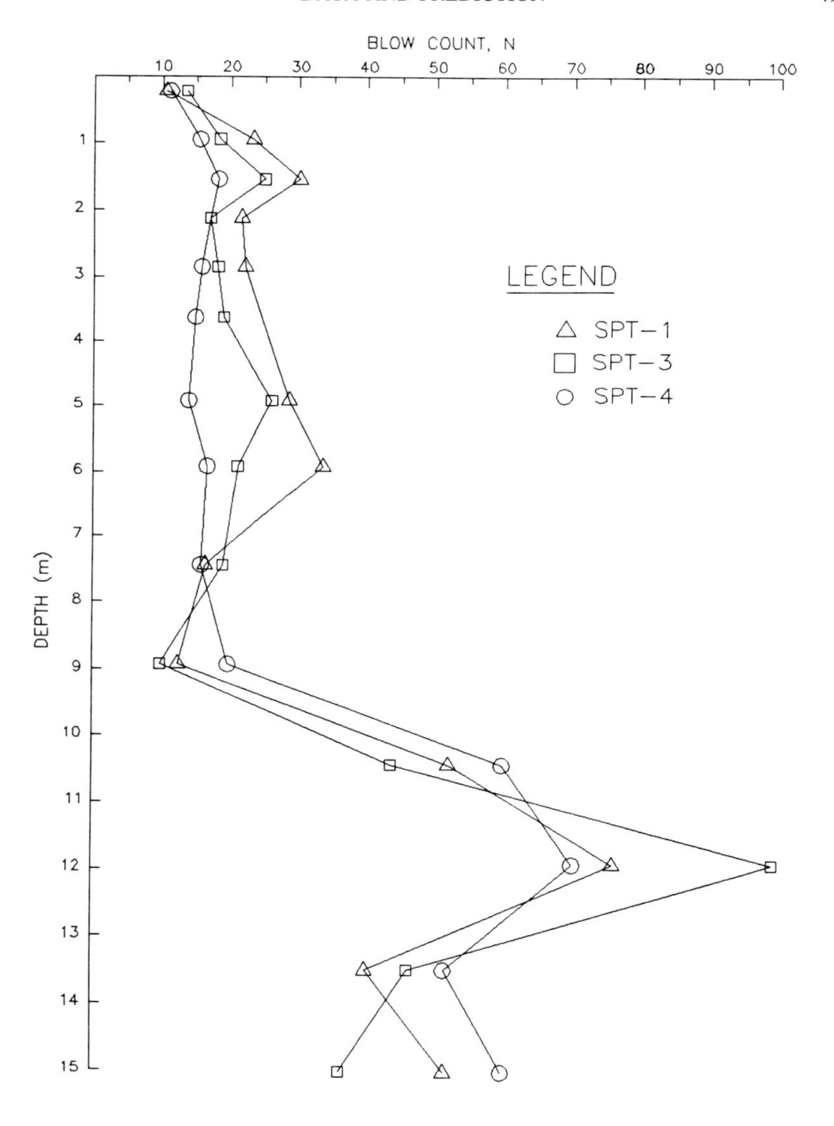

FIGURE 21
Graph of Blow Counts N Versus Depth

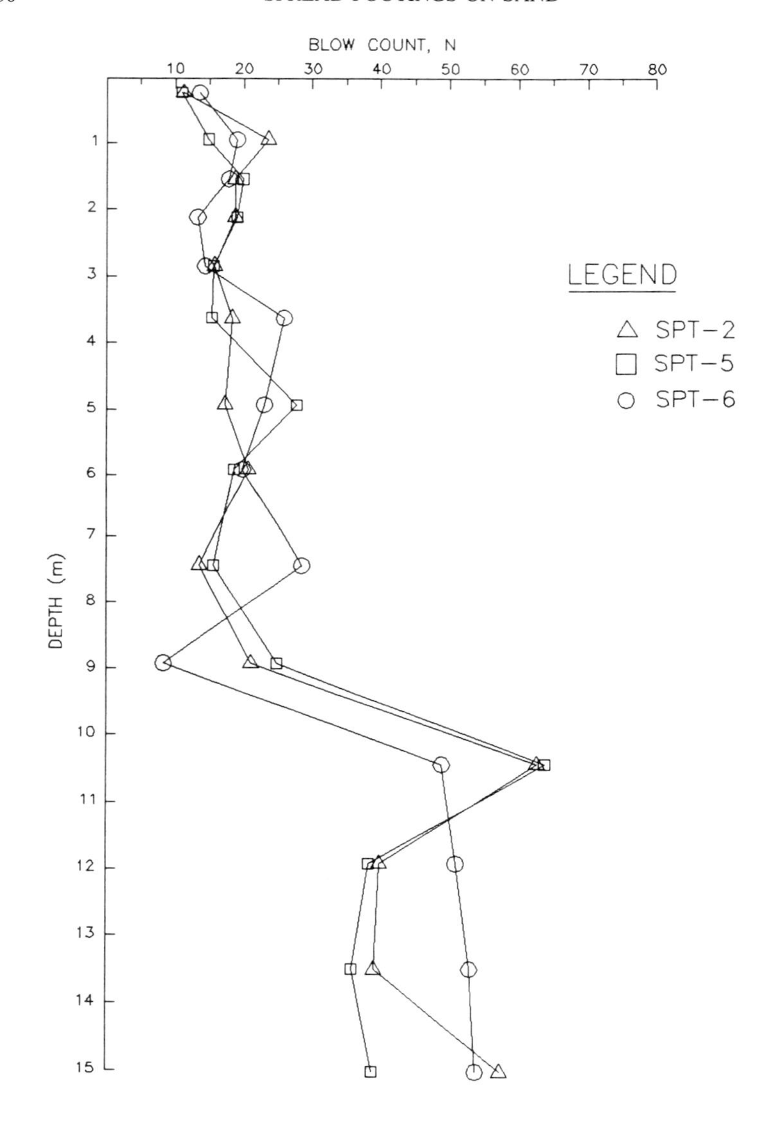

FIGURE 22
Graph of Blow Counts N Versus Depth

Test Penetration D / Rod Length R	Average Rod Top Force (kN)	Average Ram Impact Velocity (m/s)	Average Ram Kinetic Energy (kN-m)	Average Maximum Transferred Energy (FV) (kN-m)	Average System Efficiency (FV) %	Average Transferred Energy (FF) (kN-m)	Corrected Transferred Energy (ASTM D 4633-86) (kN-m)	Average System Efficiency (FF) %	Operating Speed (bpm)	SPT Blow Count bl/150 mm
D=0.8-1.2	74.3	3.0	0.28	0.24	51	0.07	0.11?	23?	48	7/10/13
R=2.4										
D=1.4-1.8	76.1	2.9	0.27	0.22	45	0.15	0.26	54	48	8/8/10
R=2.4										
D=2.0-2.4	70.8	3.0	0.28	0.23	48	0.15	0.26	54	49	6/9/9
R=4.0										
D=2.6-3.0	71.6	3.0	0.28	0.26	54	0.16	0.22	46	47	4/8/8
R=4.0										
D=3.5-4.0	66.8	2.8	0.26	0.18	38	0.19	0.22	46	50	10/10/9
R=5.5										
D=4.6-5.0	71.6	3.0	0.28	0.23	49	0.20	0.23	49	48	7/9/8
R=5.5										
D=5.6-6.1	76.5	3.0	0.27	0.23	48	0.23	0.26	53	50	7/10/11
R=7.0										
D=7.2-7.6	71.2	2.9	0.26	0.26	54	0.23	0.24	51	47	5/6/8
R=8.5										
D=8.7-9.1	73.9	3.0	0.28	0.23	49	0.24	0.24	51	48	6/8/13
R=10.1										
D=10.2-10.7	70.3	2.7	0.24	0.23	49	0.26	0.26	54	50	11/24/39
R=11.6										
D=11.7-12.2	77.0	2.8	0.24	0.24	51	0.28	0.28	60	48	11/16/24
R=13.1										
D=13.3-13.7	76.1	2.9	0.27	0.28	59	0.28	0.28	60	46	14/17/22
R=14.6										
D=14.8-15.20	75.7	2.8	0.24	0.24	51	0.27	0.27	57	50	18/25/32
R=16.2										
							E(avg) = 0.25	e(avg) = 53		

TABLE 9
Summary of Field Results for SPT Energy Measurements

3.7 PiezoCone Penetration Tests:

Electric PiezoCone Penetrometer tests were performed by Fugro-McClelland (Southwest), Inc. in Houston (Recep Yilmaz). A total of five CPT soundings were performed at the approximate locations of the footings (Figure 3) in accordance with ASTM D3441. The tests were done using a 200 kN truck which continuously pushed a standard 35.7 mm piezocone at a rate of 200 mm/sec. Side friction, tip resistance and pore pressures were read and recorded for all tests to a total depth of 16.4 m with the exceptions of soundings numbers 02 and 07. These soundings had to be stopped short of 16.4 m due to high pore pressures which threatened to blow the pore pressure transducer of the cone. The pore pressure stone was located at mid height of the cone tip.

The results of the CPT soundings can be found on figures 23 through 27 for tip resistance, friction and pore pressure.

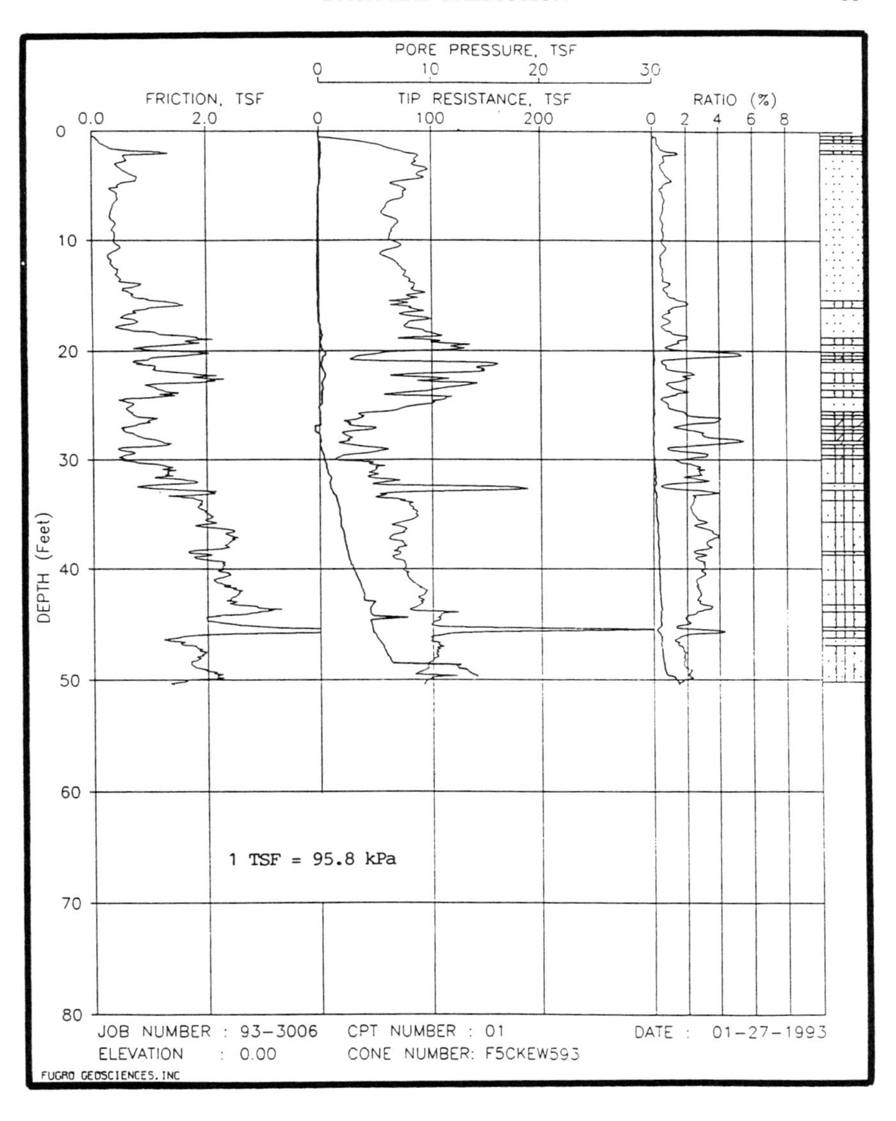

FIGURE 23
Cone Penetrometer Test Results

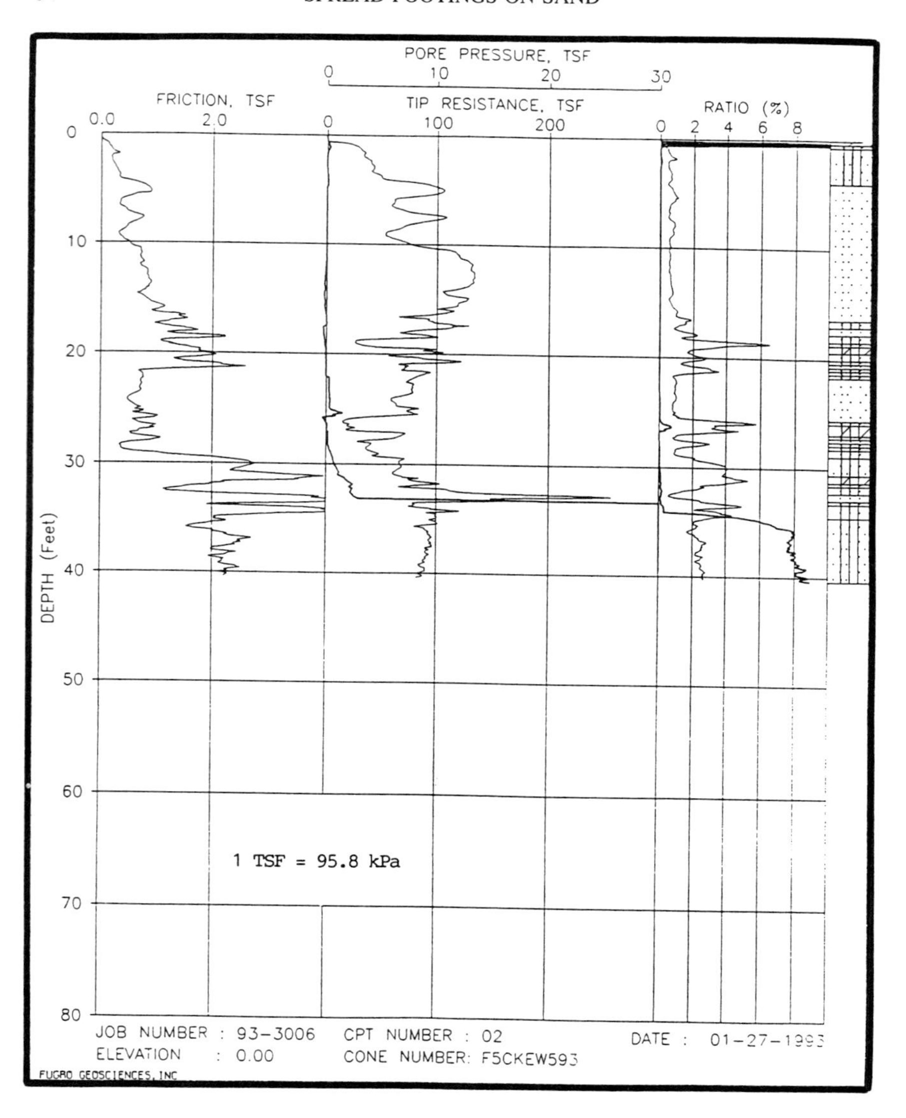

FIGURE 24
Cone Penetrometer Test Results

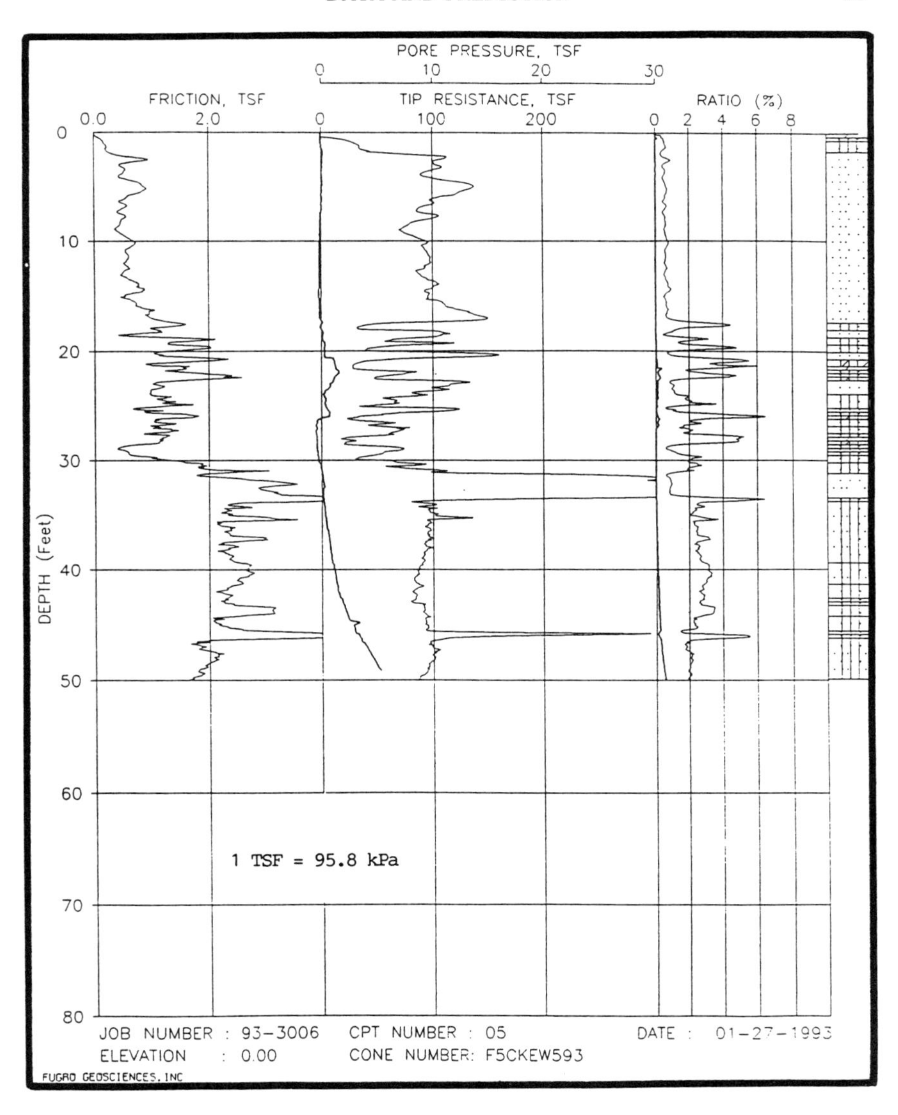

FIGURE 25
Cone Penetrometer Test Results

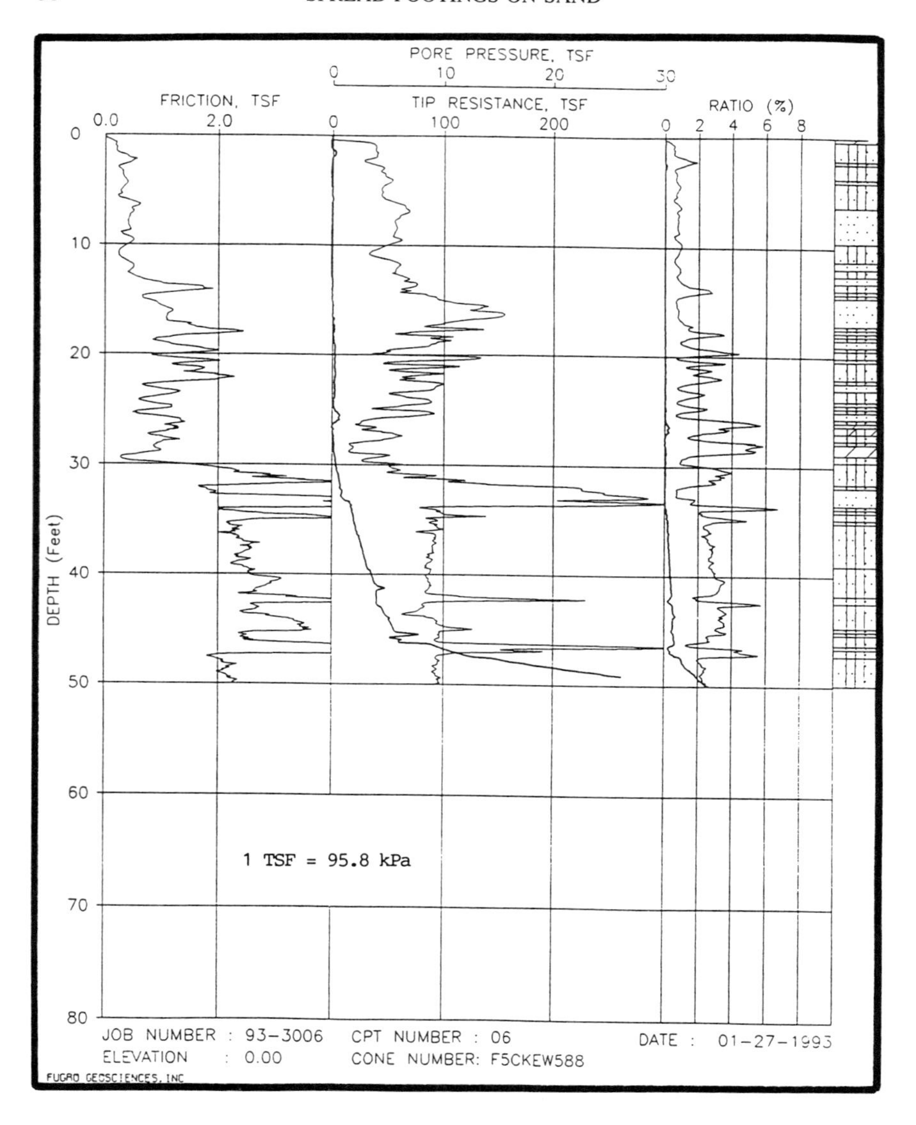

FIGURE 26
Cone Penetrometer Test Results

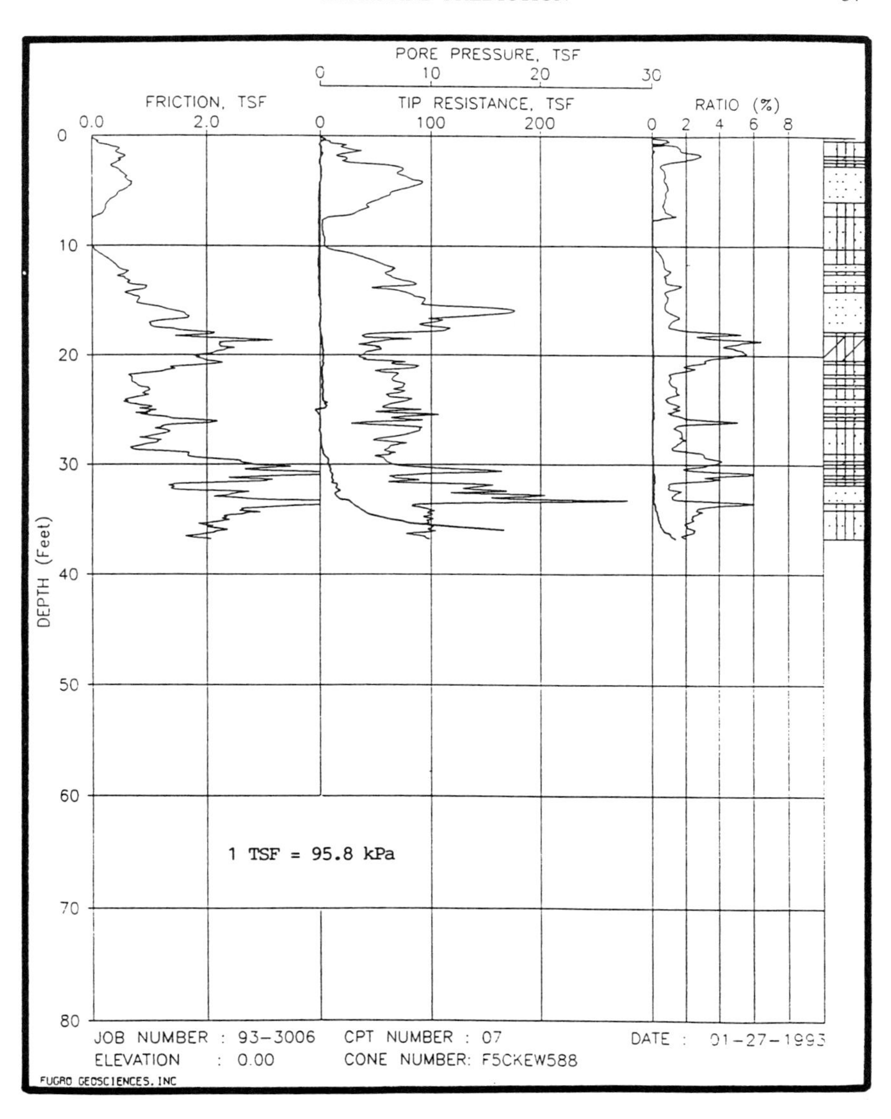

FIGURE 27
Cone Penetrometer Test Results

3.8 Pressuremeter Tests:

Pressuremeter tests were performed by Texas A&M University (Jean-Louis Briaud and Philippe Jeanjean) and Geotest Engineering (Robert Gibbens). A total of four pressuremeter borings were tested at the site all in accordance with ASTM D4719. The location of the borings can be found on Figure 3.

Borings PMT-1 and 2 were performed in two phases, the first in a hand augered hole to a depth of 4 m and the second, from 4 to 11 m, in a hole prepared by a drill rig. The hand augered holes were prepared with a 75 mm hand auger and tests were performed at 0.6, 1.2, 2.1 and 3.4 m depths. The second phase involved testing at 5.2, 7.6 and 11 m, which required the use of a drill rig. Buchanan/Soil Mechanics, Inc. did the drilling with a Failing model 1500 truck mounted rotary drilling rig. The wet rotary method was used to advance the borehole with a 75 mm wash bit. It was necessary to use drilling mud to keep the borehole from collapsing below a depth of 4 m. The pressuremeter was attached to the drill string and lowered in the open hole to the desired depth.

Borings PMT-3 and 4 were performed after PMT-1 and 2 at locations shown on Figure 3. These tests were performed using a 75 mm hand auger at depths of 1.3, 2.1 and 3.0 m.

All the results for the pressuremeter tests can be found on Figures 28 through 31 in the form of limit pressure versus depth, modulus versus depth and reload modulus versus depth and on Table 10. Figure 32 shows an example of the pressuremeter curve for pmt-2.

During the pressuremeter test, some special effects were investigated including the influence of the strain level, the stress level and the pressure duration on the modulus.

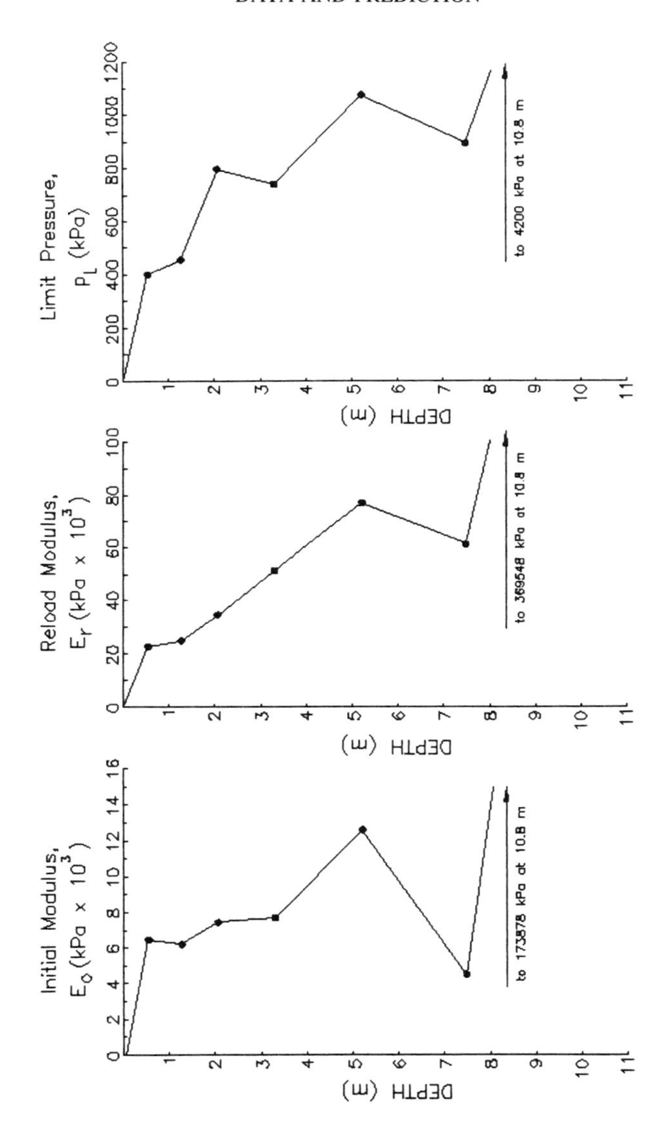

FIGURE 28
Pressuremeter Test Results for PMT-1

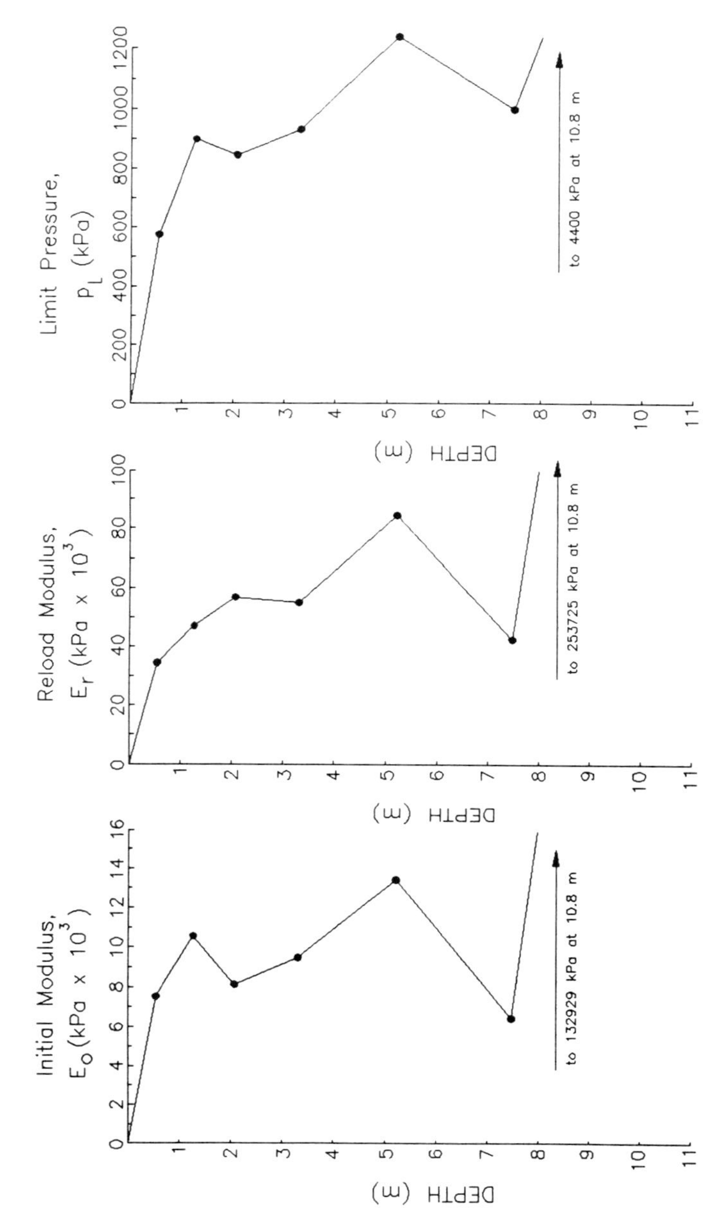

FIGURE 29
Pressuremeter Test Results for PMT-2

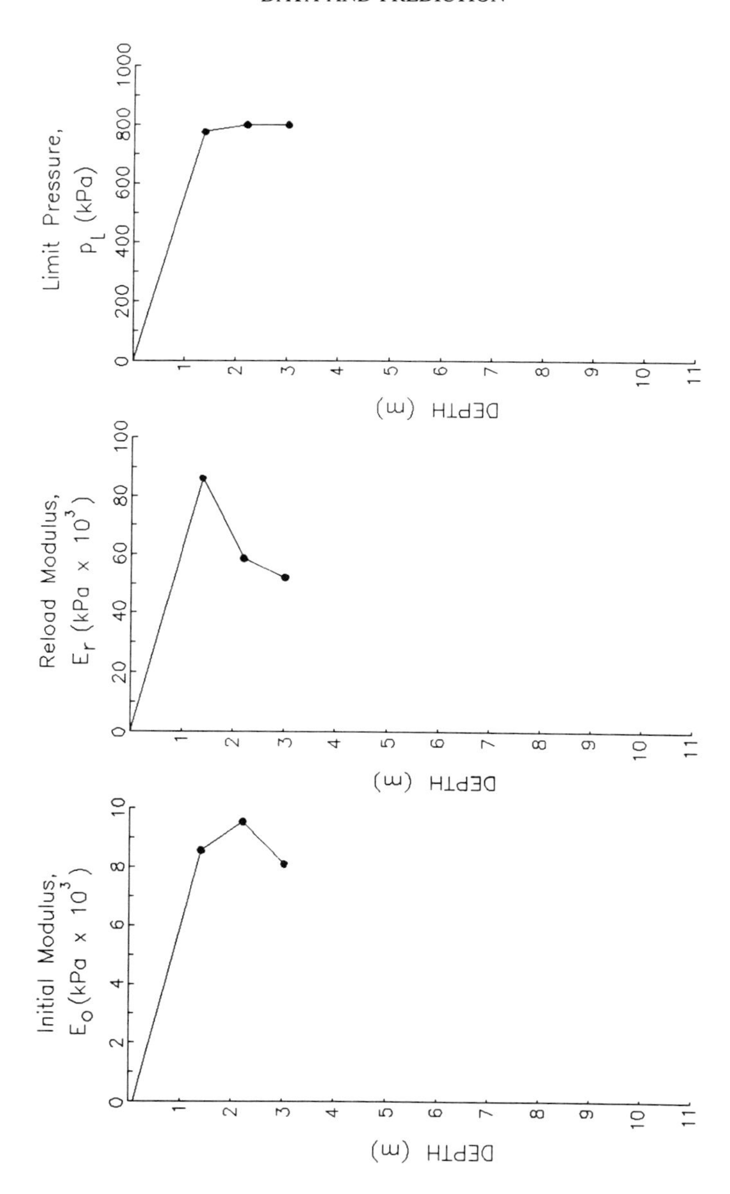

FIGURE 30
Pressuremeter Test Results for PMT-3

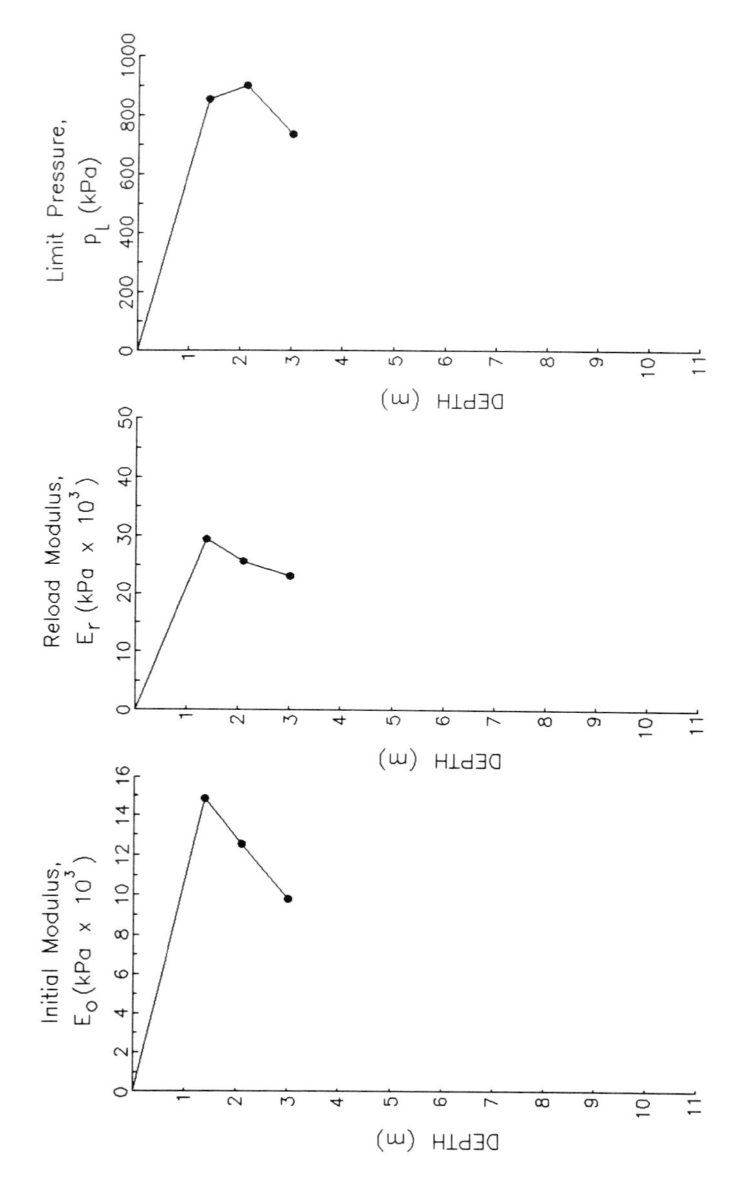

FIGURE 31
Pressuremeter Test Results for PMT-4

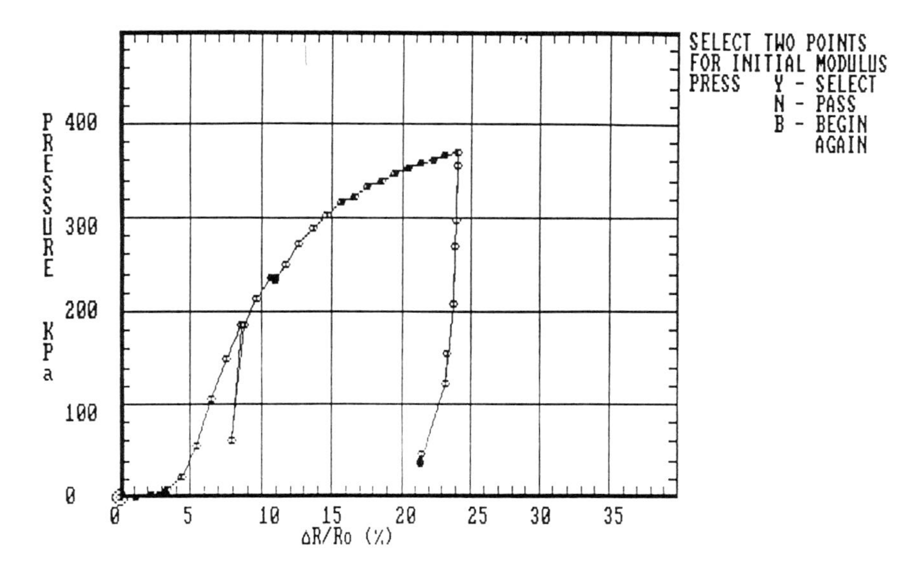

FIGURE 32
Sample Pressuremeter Curve for PMT-2

DEPTH (m)	E_0 (kPa)	E_R (kPa)	p_l (kPa)
pmt-1			
0.6	6421	21904	400
1.2	6270	24335	460
2.1	7443	33903	800
3.4	7746	51428	740
5.2	12595	78108	1100
7.6	4619	61099	900
11.0	173878	369548	4200
pmt-2			
0.6	7621	34336	580
1.2	10558	47885	900
2.1	8075	57779	840
3.4	9490	55513	920
5.2	13365	84967	1250
7.6	6387	41970	1000
11.0	132929	253725	4400

DEPTH (m)	E_0 (kPa)	E_R (kPa)	p_l (kPa)
pmt-3			
1.3	8563	87760	780
2.1	9663	59289	800
3.0	8182	52085	800
pmt-4			
1.3	14868	29572	850
2.1	12373	25450	900
3.0	9862	23089	720

TABLE 10
Tabulated Pressuremeter Results

3.9 Cross-Hole Wave Tests

Cross-hole wave tests were performed by Derek Morris, Tony Yen and John Delphia of Texas A&M University. The tests were performed in accordance with ASTM D4428, using an impulsive down-hole energy source. This consisted of an expanding down-hole borehole clamp that could be given a sharp vertical impact from above, via a hammer (length of pipe) sliding on the galvanized pipe that held the borehole clamp. The times of arrival were measured with two down-hole geophones, consisting of two OYO SCM-730 8 Hz geophones, suitably weighted and equipped with a pneumatic bladder borehole clamp.

The tests were performed in the grouted inclinometer casings at the site, with one casing housing the down-hole hammer and two others housing the geophones. Two series of in-line tests were performed at the site. The first line of tests consisted of the three casings representing the North-South direction (cht-1, 2 and 3 on Figure 3). The down-hole hammer was placed in cht-3, with the two geophones placed in cht-1 and 2. The second series of tests, representing the East-West direction, consisted of the down-hole hammer in cht-4 and the two geophones in cht-2 and 5 (Figure 3). Inclinometer tests were performed on all casings for an accurate determination of the horizontal distance between geophones.

The results of the cross-hole shear wave test are presented in Table 11.

North-South direction between holes cht-2 and cht-1 (Figure 3), nominal surface spacing = 2.415 m

Depth (m)	Time Difference (ms)	Corrected Separation (m)	Shear Wave Velocity (m/s)	Shear Modulus (MPa)
2	10	2.398	240	104
4	8	2.397	300	162
6	8.5	2.391	281	142
8	12	2.383	199	71
10	10	2.380	238	102

East-West direction between holes cht-2 and cht-5 (Figure 3), nominal surface spacing = 1.924 m

Depth (m)	Time Difference (ms)	Corrected Separation (m)	Shear Wave Velocity (m/s)	Shear Modulus (MPa)
2	9.5	1.918	202	73
4	9	1.902	211	80
6	9	1.887	210	79
8	11	1.865	170	52
10	8	1.839	230	95

TABLE 11
Tabulated Cross-Hole Test Results

3.10 Dilatometer Tests

Dilatometer Tests were performed by the University of L'Aquila (Maurizio Calabrese, Silvano Marchetti) and Geotest Engineering (Robert Gibbens). The tests were performed using a Failing Model 1500 truck mounted rotary drilling rig with a 121 mm wash bit used to advance the borehole together with drilling mud to keep the borehole from collapsing.

The drill rig used for the Dilatometer tests was only capable of pushing the DMT blade about 2 m at a time. After the DMT blade was pushed for the first 2 m (reading every 0.2 m), it was removed from the borehole and the borehole was augered to a depth of 2 m. The DMT blade was then reinserted into the borehole at the end of the drill string and pushed another 2 m. This process was repeated until the rig could no longer push the DMT blade.

Two DMT soundings (dmt-1 & 2 on Figure 3) were performed by Mr. Calabrese during his visit to the site in April 1993. Another DMT sounding (dmt-3 on Figure 3) was performed by Robert Gibbens in May 1993 with axial thrust measurements during this test. A graph of axial thrust versus depth can be found on Figure 36.

The results of all DMT tests can be found in Figures 33 through 35 and on Tables 12 through 14.

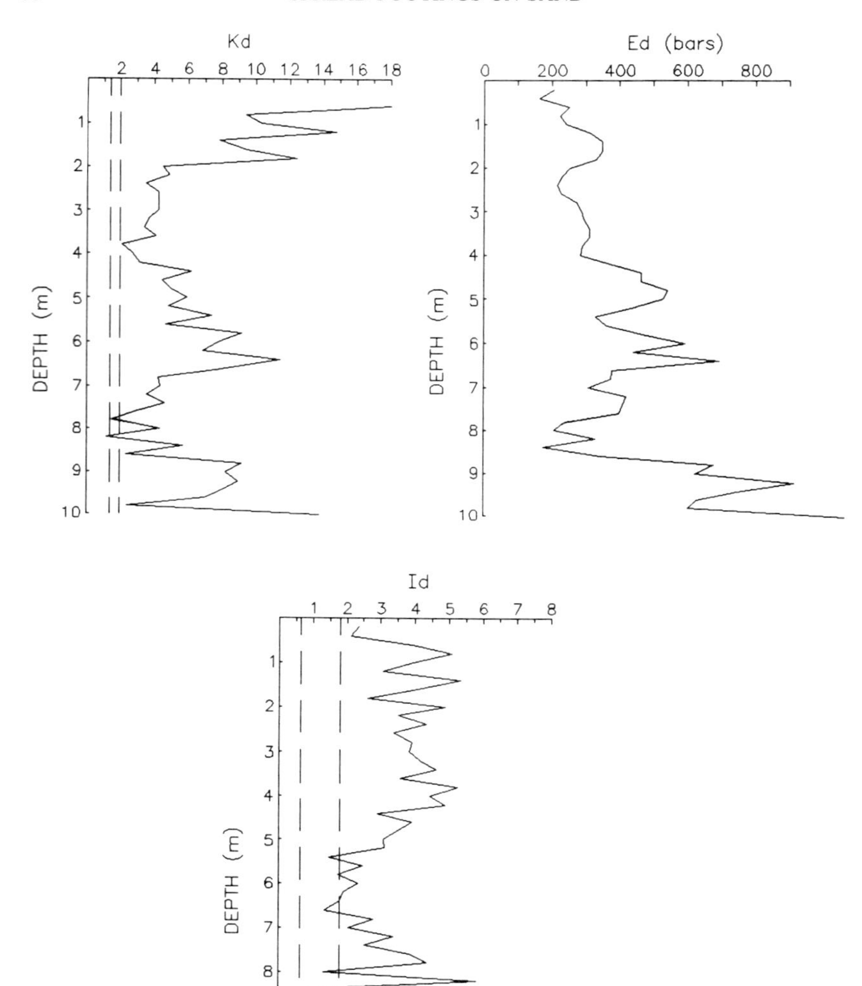

FIGURE 33
DMT-1 Test Results w/o Axial Thrust Measurements

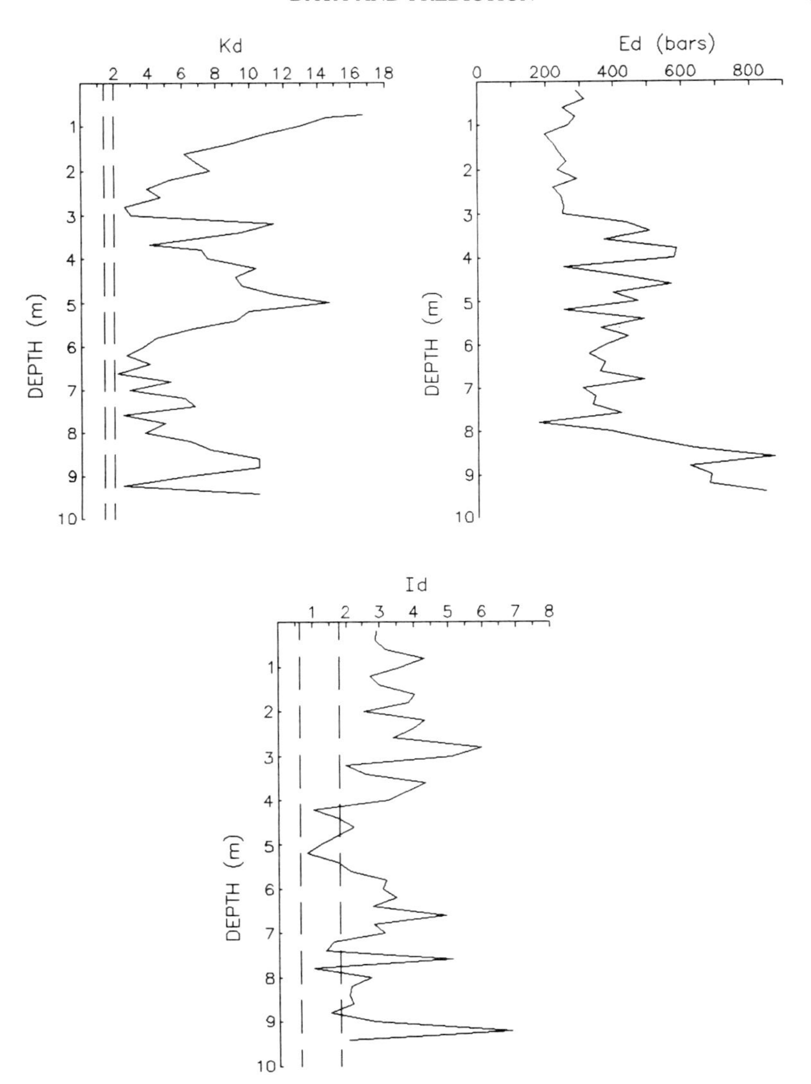

FIGURE 34
DMT-2 Test Results w/o Axial Thrust Measurements

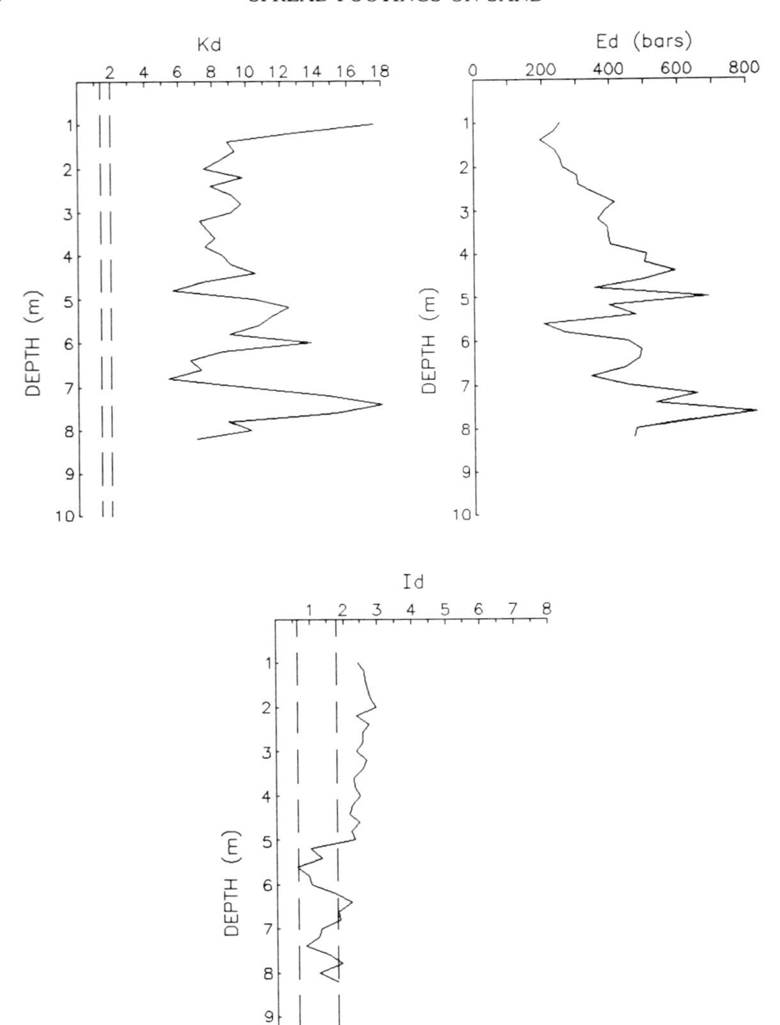

FIGURE 35
DMT-3 Test Results w/Axial Thrust Measurements

DMT-1 w/o AXIAL THRUST MEASUREMENTS
APRIL 8, 1993

(m)	EFFECTIVE STRESS (bars)	Po (bars)	P1 (bars)	U (bars)	Ed (bars)	Id	Kd
0.20	0.03	2.4	8.3	0.00	204	2.44	70.7
0.40	0.07	2.1	6.9	0.00	164	2.21	31.3
0.60	0.10	1.8	9.2	0.00	254	3.97	18.0
0.80	0.14	1.3	7.8	0.00	226	5.10	9.4
1.00	0.17	1.7	8.7	0.00	242	3.98	10.3
1.20	0.20	3.0	12.2	0.00	318	3.04	14.8
1.40	0.24	1.9	12.0	0.00	351	5.39	7.9
1.60	0.27	2.6	12.7	0.00	351	3.89	9.5
1.80	0.31	3.8	13.6	0.00	338	2.57	12.4
2.00	0.34	1.5	9.0	0.00	258	4.87	4.5
2.20	0.37	1.9	8.4	0.00	226	3.52	4.9
2.40	0.41	1.4	7.7	0.00	217	4.34	3.5
2.60	0.44	1.9	8.3	0.00	224	3.44	4.2
2.80	0.48	2.0	10.0	0.00	278	3.97	4.2
3.00	0.51	2.1	10.4	0.00	286	3.88	4.2
3.20	0.54	2.0	10.5	0.00	293	4.18	3.7
3.40	0.58	1.9	11.0	0.00	314	4.64	3.4
3.60	0.61	2.5	11.5	0.00	310	3.57	4.1
3.80	0.65	1.6	10.0	0.00	291	5.25	2.5
4.00	0.68	1.8	10.0	0.00	284	4.48	2.7
4.20	0.72	2.2	13.0	0.00	375	4.89	3.1
4.40	0.75	4.7	18.1	0.00	466	2.88	6.2
4.60	0.78	3.4	16.9	0.00	468	3.94	4.4
4.80	0.82	4.1	19.7	0.00	541	3.78	5.0
5.00	0.84	5.0	20.3	0.01	530	3.04	6.0
5.20	0.86	4.2	17.1	0.03	446	3.05	4.9
5.40	0.87	6.5	16.1	0.05	334	1.50	7.4
5.60	0.89	4.3	14.7	0.07	362	2.49	4.7
5.80	0.90	8.3	22.0	0.09	474	1.66	9.2
6.00	0.91	7.3	24.3	0.11	591	2.37	7.8
6.20	0.93	6.5	18.7	0.13	422	1.91	6.9
6.40	0.94	11.0	31.0	0.15	694	1.85	11.5
6.60	0.96	7.9	18.8	0.17	380	1.42	8.0
6.80	0.97	4.3	15.2	0.19	379	2.68	4.2
7.00	0.99	4.4	13.3	0.21	308	2.09	4.3
7.20	1.00	3.8	15.9	0.23	420	3.41	3.6
7.40	1.01	4.9	16.8	0.25	412	2.55	4.6
7.60	1.03	3.3	14.8	0.26	399	3.80	2.9
7.80	1.04	1.9	8.8	0.28	241	4.41	1.5
8.00	1.06	4.9	11.1	0.30	215	1.35	4.3
8.20	1.07	2.0	11.7	0.32	337	5.79	1.6
8.40	1.09	6.6	11.8	0.34	179	0.82	5.8
8.60	1.10	2.9	12.9	0.36	345	3.85	2.3
8.80	1.12	10.7	30.3	0.38	680	1.90	9.3
9.00	1.13	9.7	27.6	0.40	622	1.93	8.2
9.20	1.14	10.8	37.3	0.42	920	2.56	9.0
9.40	1.16	9.8	31.5	0.44	753	2.32	8.1
9.60	1.17	8.7	26.9	0.46	632	2.21	7.0
9.80	1.19	3.2	20.5	0.48	602	6.46	2.3
10.00	1.20	17.2	49.3	0.50	1113	1.92	13.9

Po = Corrected A reading
P1 = Corrected B reading
U = Pore pressure
Ed = Dilatometer modulus
Id = Material index
Kd = Horizontal stress index

TABLE 12
Tabulated DMT Test Results for DMT-1

DMT-2 w/o AXIAL THRUST MEASUREMENTS
APRIL 13, 1993

(m)	EFFECTIVE STRESS (bars)	Po (bars)	P1 (bars)	U (bars)	Ed (bars)	Id	Kd
0.20	0.03	2.9	11.4	0.00	295	2.94	85.0
0.40	0.07	3.1	12.0	0.00	308	2.82	46.1
0.60	0.10	2.2	9.4	0.00	248	3.18	22.0
0.80	0.14	1.9	10.4	0.00	293	4.34	14.3
1.00	0.17	2.2	10.1	0.00	274	3.57	13.0
1.20	0.20	2.1	7.9	0.00	199	2.73	10.3
1.40	0.24	2.1	8.5	0.00	223	3.04	8.9
1.60	0.27	1.7	8.4	0.00	234	4.06	6.1
1.80	0.31	2.0	9.5	0.00	261	3.83	6.4
2.00	0.34	2.7	9.4	0.00	235	2.54	7.8
2.20	0.37	2.0	10.4	0.00	293	4.31	5.2
2.40	0.41	1.6	8.1	0.00	224	3.96	4.0
2.60	0.44	2.1	9.2	0.00	245	3.39	4.7
2.80	0.48	1.3	8.8	0.00	263	5.96	2.7
3.00	0.51	1.5	9.3	0.00	269	5.12	3.0
3.20	0.54	6.2	18.6	0.00	431	2.01	11.4
3.40	0.58	5.5	20.1	0.00	504	2.62	9.6
3.60	0.61	2.4	13.3	0.00	377	4.44	4.0
3.80	0.65	4.6	21.8	0.00	597	3.74	7.1
4.00	0.68	5.2	22.1	0.00	587	3.25	7.6
4.20	0.72	7.3	14.9	0.00	262	1.03	10.3
4.40	0.75	6.9	19.0	0.00	419	1.74	9.2
4.60	0.78	7.6	24.3	0.00	581	2.22	9.6
4.80	0.82	9.4	20.9	0.00	399	1.22	11.5
5.00	0.84	12.5	26.3	0.01	479	1.11	14.8
5.20	0.86	8.6	16.1	0.03	258	0.87	10.0
5.40	0.87	8.0	22.2	0.05	491	1.77	9.2
5.60	0.89	5.7	17.6	0.07	412	2.11	6.4
5.80	0.90	4.1	16.8	0.09	442	3.22	4.4
6.00	0.91	3.6	14.3	0.11	372	3.10	3.8
6.20	0.93	2.8	11.9	0.13	317	3.48	2.8
6.40	0.94	4.0	14.7	0.15	372	2.79	4.1
6.60	0.96	2.2	12.4	0.17	353	4.97	2.1
6.80	0.97	5.3	19.6	0.19	497	2.81	5.2
7.00	0.99	3.1	12.0	0.21	310	3.14	2.9
7.20	1.00	6.4	16.2	0.23	340	1.59	6.2
7.40	1.01	7.1	16.8	0.25	337	1.42	6.8
7.60	1.03	2.6	14.7	0.26	419	5.14	2.3
7.80	1.04	5.5	10.8	0.28	184	1.01	5.0
8.00	1.06	4.4	15.7	0.30	391	2.74	3.9
8.20	1.07	7.1	21.5	0.32	501	2.14	6.3
8.40	1.09	8.9	26.9	0.34	626	2.12	7.8
8.60	1.10	11.9	37.3	0.36	880	2.19	10.5
8.80	1.12	12.2	30.1	0.38	622	1.52	10.6
9.00	1.13	7.3	27.3	0.40	694	2.89	6.1
9.20	1.14	3.3	22.9	0.42	682	6.94	2.5
9.40	1.16	12.6	37.1	0.44	851	2.02	10.5

Po = Corrected A reading
P1 = Corrected B reading
U = Pore pressure
Ed = Dilatometer modulus
Id = Material index
Kd = Horizontal stress index

TABLE 13
Tabulated DMT Test Results for DMT-2

DMT-3 w/AXIAL THRUST MEASUREMENTS
MAY 27, 1993

DEPTH (m)	EFFECTIVE STRESS (bars)	Po (bars)	P1 (bars)	U (bars)	Ed (bars)	Id	Kd	THRUST (kN)
1.00	0.17	3.00	10.40	0.00	256.87	2.47	17.60	20.0
1.20	0.20	2.62	9.48	0.00	237.92	2.61	12.84	20.0
1.40	0.24	2.12	7.75	0.00	195.29	2.65	8.90	18.6
1.60	0.27	2.52	9.40	0.00	238.65	2.73	9.26	19.3
1.80	0.31	2.60	9.90	0.00	253.22	2.80	8.49	20.0
2.00	0.34	2.54	10.10	0.00	262.33	2.98	7.46	21.7
2.20	0.37	3.68	12.51	0.00	306.42	2.40	9.82	23.3
2.40	0.41	3.23	12.10	0.00	307.88	2.75	7.90	23.6
2.60	0.44	4.04	14.29	0.00	355.61	2.54	9.13	25.0
2.80	0.48	4.60	16.50	0.00	412.81	2.58	9.65	25.5
3.00	0.51	4.62	15.70	0.00	384.39	2.40	9.05	27.4
3.20	0.54	3.95	14.50	0.00	366.17	2.67	7.24	27.0
3.40	0.58	4.46	15.89	0.00	396.78	2.57	7.69	25.2
3.60	0.61	5.00	16.50	0.00	398.96	2.30	8.16	25.1
3.80	0.65	4.94	16.53	0.00	402.24	2.35	7.63	25.4
4.00	0.68	5.85	20.50	0.00	508.27	2.50	8.59	28.0
4.20	0.72	6.51	21.10	0.00	506.45	2.24	9.09	32.3
4.40	0.75	7.83	24.90	0.00	592.43	2.18	10.45	37.4
4.60	0.78	5.81	20.30	0.00	502.80	2.49	7.42	28.8
4.80	0.82	4.57	14.70	0.00	351.60	2.22	5.59	28.6
5.00	0.84	8.55	28.50	0.01	692.27	2.34	10.15	33.4
5.20	0.86	10.74	22.25	0.03	399.33	1.07	12.51	31.5
5.40	0.87	10.05	23.70	0.05	473.66	1.36	11.49	30.7
5.60	0.89	9.52	15.40	0.07	204.04	0.62	10.68	25.8
5.80	0.90	8.18	15.90	0.09	267.80	0.95	9.00	34.5
6.00	0.91	12.68	25.70	0.11	451.79	1.04	13.76	34.1
6.20	0.93	8.13	22.29	0.13	491.51	1.77	8.62	38.2
6.40	0.94	6.48	20.51	0.15	486.77	2.21	6.72	31.8
6.60	0.96	7.14	19.90	0.17	442.69	1.83	7.29	33.9
6.80	0.97	5.43	15.40	0.19	346.13	1.90	5.39	30.5
7.00	0.99	10.17	23.40	0.21	459.08	1.33	10.10	30.8
7.20	1.00	15.10	34.00	0.23	655.83	1.27	14.87	29.2
7.40	1.01	18.57	34.00	0.25	535.59	0.84	18.05	33.1
7.60	1.03	15.96	39.90	0.26	830.72	1.53	15.25	40.8
7.80	1.04	9.63	27.90	0.28	633.97	1.95	8.95	45.1
8.00	1.06	11.15	24.90	0.30	477.30	1.27	10.24	41.2
8.20	1.07	7.85	21.50	0.32	473.66	1.81	7.02	31.6

Po = Corrected A reading
P1 = Corrected B reading
U = Pore pressure
Ed = Dilatometer modulus
Id = Material index
Kd = Horizontal stress index

TABLE 14
Tabulated DMT Test Results for DMT-3

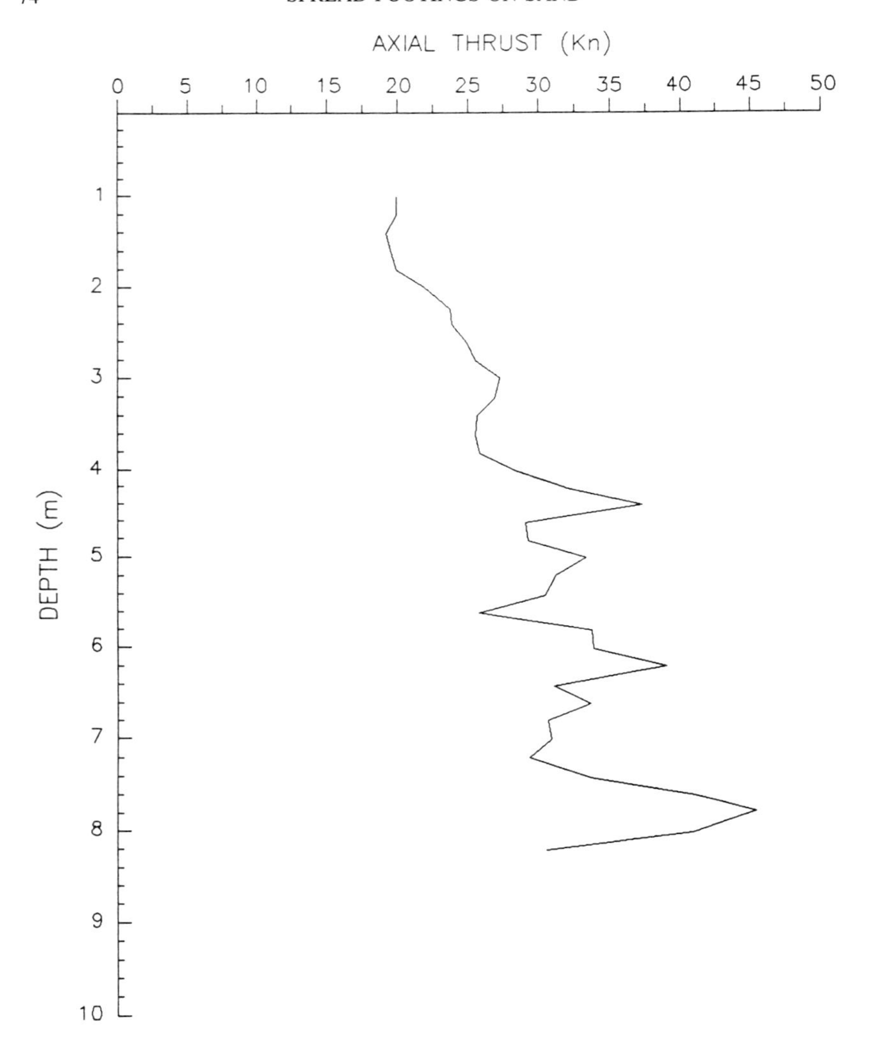

FIGURE 36
Axial Thrust Versus Depth for DMT-3

3.11 Borehole Shear Tests

Borehole shear tests were performed at the site by Allan Lutenneger and Don Degroot of the University of Massachusetts at Amherst and Michael Adams of the Federal Highway Administration in Washington. A total of three borehole shear test borings were performed at locations found on Figure 3. The holes for the borehole shear tests were drilled with a 75 mm hand auger. At each prescribed depth a borehole shear test was run at 6 different normal stresses. Then, the hole was hand augered to a deeper depth where the same test procedure was followed.

The tabulated results of the borehole shear tests can be found on Table 15. The results can also be seen in a graph of ϕ versus depth on Figure 37.

Boring No.	Test No.	Depth (m)	ϕ (deg.)	c (N/m^2)	R^2
BST-1	1-1	0.61	33.6	-69.7	0.999
	1-2	1.2	33.3	-34.8	1.000
	1-3	1.8	33.2	-34.8	1.000
	1-4	2.4	32.7	-34.8	0.999
	1-5	3.0	36.1	0	1.000
	1-6	3.7	31.6	0	1.000
	1-7	4.3	31.0	104.5	0.998
	1-8	4.6	31.3	0	0.999
BST-2	2-1	0.61	38.1	-104.5	0.989
	2-2	1.2	37.8	34.8	0.999
	2-3	1.8	32.8	-104.5	0.999
	2-4	2.4	30.3	34.8	0.998
	2-5	3.0	27.6	104.5	0.996
	2-6	3.7	25.0	174.2	0.995
	2-7	4.3	24.3	209.0	0.998
	2-8	4.9	26.1	69.7	1.000
BST-3	3-1	0.61	33.2	34.8	0.999
	3-2	1.2	33.9	-104.5	1.000
	3-3	1.8	33.6	-243.8	0.999
	3-4	2.4	29.2	139.3	1.000
	3-5	3.0	29.4	69.7	1.000
	3-6	3.7	27.0	209.0	0.997
	3-7	4.3	31.1	69.7	0.999

TABLE 15
Tabulated Borehole Shear Test Results

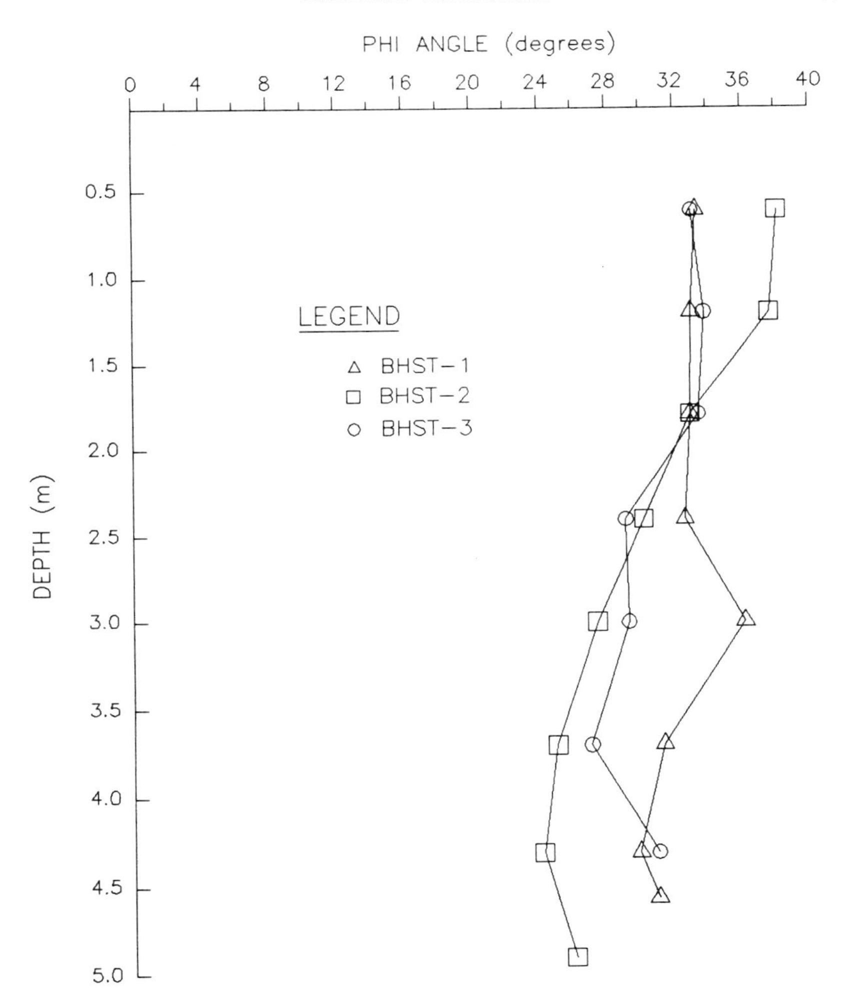

FIGURE 37

Graph of ϕ Versus Depth from Borehole Shear Results

3.12 Step Blade Tests

The Step Blade test was performed at the site by Allan Lutenneger and Don Degroot of the University of Massachusetts at Amherst. Buchanan/Soil Mechanics, Inc. performed the drilling for the test with a Failing model 1500 truck mounted rotary drilling rig. A 121 mm wash bit was used to advance the borehole along with drilling mud to keep the borehole from collapsing. Tests were performed every 0.76 m to a total depth of 6.1 m. The location of the test holes can be found in Figure 3.

The test was performed by pushing the step blade at the bottom of the open borehole. The step blade is made of four blades of increasing thickness: t_1, t_2, t_3 and t_4 (Figure 38). The t_1 thick blade was pushed in the soil at the bottom of the borehole. A reading was taken on the cell of the t_1 thick blade. The t_2 thick blade was then pushed in the soil so that cell 2 was at the depth where cell 1 used to be. A reading was taken on cell 2. The same process was repeated for cells 3 and 4. The results can be found in Table 16.

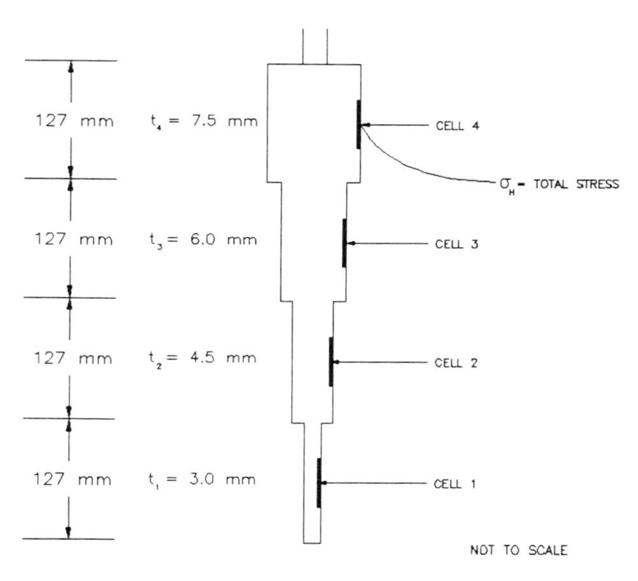

FIGURE 38
Step Blade Cross Section

Depth (m)	CELL PRESSURE (kPa)			
	Cell-1	Cell-2	Cell-3	Cell-4
0.76	69	62	18	35
1.5	104	86	93	80
2.2	93	38	62	86
3.0	93	69	76	90
3.8	76	55	49	66
4.6	76	36	NR	73
5.3	80	35	35	45
6.1	80	73	76	40

TABLE 16
FHWA Step Blade Test Results

4. PREDICTION REQUEST

The load test results for each footing can be described by the generic stepped line in Figure 39. The 30 minute settlement curve will be obtained by joining the points corresponding to the 30 minute readings. The 1 minute load settlement curve will be obtained by joining the points corresponding to the 1 minute readings.

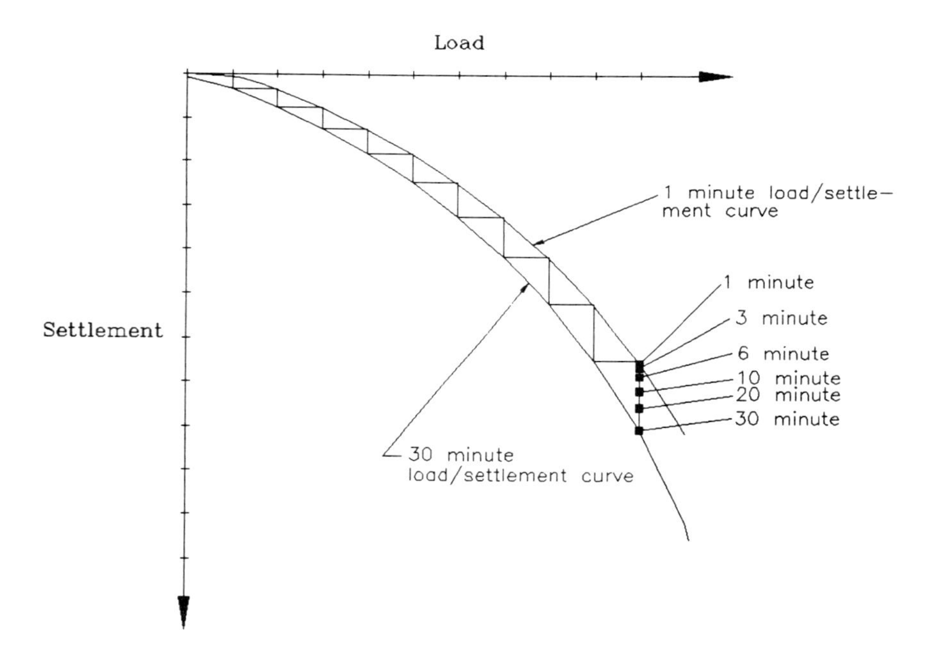

FIGURE 39
Load/Settlement Curve

The participants to the prediction event are requested to predict the following quantities:

For each of the five footings:

1. What will be the load measured in the load test at a settlement of 25 mm on the 30 minute load settlement curve (Figure 39)?
2. What will be the load measured in the load test at a settlement of 150 mm on the 30 minute load settlement curve (Figure 39)?

For the 3 x 3 m footing:

3. For the load corresponding to a settlement of 25 mm on the 30 minute load settlement curve, what will be the amount of creep settlement which will take place between the 1 minute reading and the 30 minute reading (Figure 40)?
4. For the load corresponding to a settlement of 25 mm on the 30 minute load settlement curve, what will be the settlement in the year 2014 if that load is applied for the next 20 years?

The participants are requested to send their predictions in the form of a paper following the ASCE Journal of Geotechnical Engineering guidelines. The paper must include the table below; the paper is to be on 216 mm x 279 mm paper; the text must be single-spaced; the margin must be 25 mm all around; references must be given for any method used; the total number of pages must be less than or equal to 4 including figures, tables and references; please give your answers in SI units; the paper should have a title with the name of the authors and their address as shown; the content should give an explanation of how the predictions were reached and any other comments. The summary predictions must be presented in a table similar to Table 17.

The papers will be collected in a prediction volume together with a summary analysis of the comparison between predicted and measured results, and with this report as an appendix. The prediction volume will be available at the conference.

	Footing 1 3 m x 3 m	Footing 2 1.5 m x 1.5 m	Footing 3 3 m x 3 m	Footing 4 2 m x 2 m	Footing 5 1 m x 1 m
Load for 25 mm of settlement Q25 on the 30 minute load settlement curve (Kn)					
Load for 150 mm of settlement Q150 on the 30 minute load settlement curve (Kn)					
Creep settlement between 1 minute and 30 minute for Q25, Δs (mm)					
Settlement in the year 2014 under Q25 (mm)					

TABLE 17
Prediction Summary Format

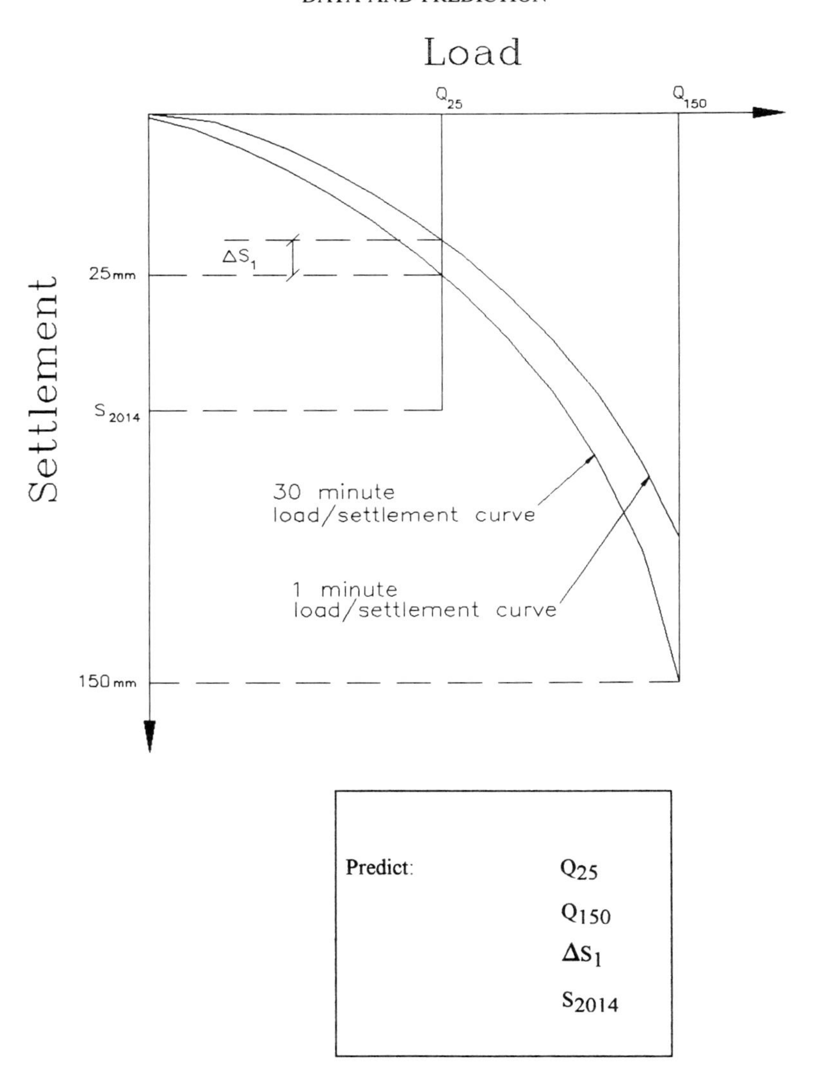

Predict:	Q_{25}
	Q_{150}
	ΔS_1
	S_{2014}

FIGURE 40
Load/Settlement Curve for Prediction Request Explaination

One original and two copies of the paper must be sent to the address below by December 1, 1993.

Professor Jean-Louis Briaud
Department of Civil Engineering
Texas A&M University
College Station, Texas 77843-3136
U.S.A.

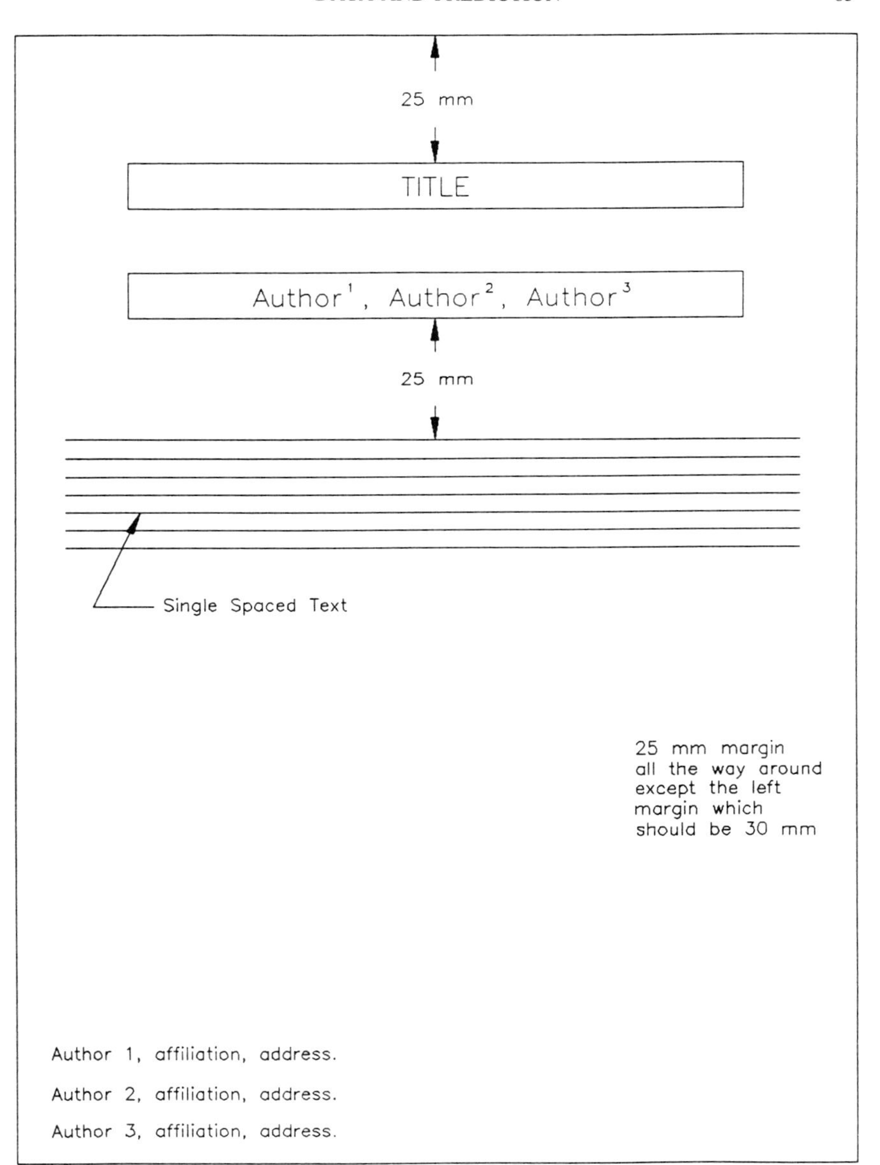
25 mm
TITLE
Author[1], Author[2], Author[3]
25 mm
Single Spaced Text
25 mm margin
all the way around
except the left
margin which
should be 30 mm
Author 1, affiliation, address.
Author 2, affiliation, address.
Author 3, affiliation, address.

ADDENDUMS

TO THE

PREDICTION PACKAGE

SETTLEMENT '94
GEOTECHNICAL DIVISION

345 East 47th Street
New York, NY 10017
(212) 705-7496

Address reply to: Dr. Jean-Louis Briaud
Geotechnical Engineering & Surveying Area
Civil Engineering Dept.
Texas A&M University
College Station, TX 77843-3136, U.S.A.

September 1, 1993

Dear Predictor:

Below is an errata and some comments for the prediction package that we sent to you a few weeks ago. We offer our apologies for any inconvenience we may have created.

1. On page 9 of the prediction package, it was stated that visual classifications were performed to a depth of 16.5 m and that the Atterberg limit sample was taken at a depth of 16.4 m. The correct maximum depth of visual classification and the Atterberg limit sample should be 15.2 m, the bottom of the borehole.

2. On page 22 of the prediction package, in table 7 listing Sample Depth, Confining Pressure and Deviatoric Stress at Failure, in the row of 3.0 m depth and 345 kPa Confining Pressure, the Deviatoric Stress at Failure should read 1030 instead of 103.

   On page 23 of the prediction package, the legend should read Stress Strain Curve for 3.0 m sample instead of Stress Strain Curve for 0.6 m sample. Also on that page the 0.34×10^5 Pa data legend needs to be switched with the 3.45×10^5 Pa data legend.

   On page 24 of the prediction package, the legend should read Volume Change Curve for 3.0 m sample instead of Volume Change Curve for 0.6 m sample.

   On page 25 of the prediction package, the legend should read Stress Strain Curve for 0.6 m sample instead of Stress Strain Curve for 3.0 m sample.

   On page 26 of the prediction package, the legend should read Volume Change Curve for 0.6 m sample instead of Volume Change Curve for 3.0 m sample.

3. The scale for the pore pressure ratio found on the right hand side of the CPT graphs on pages 40 through 44 needs to be modified. The values for the pore pressure ratio should be multiplied by a factor of 10. Instead of having the scale 0-10 percent, the scale should be 0-100 percent. The friction ratio scale is correct, however. The friction ratio is represented by the larger percentage line. The pore pressure ratio is the pore pressure, divided by the point resistance q_c.

We wish to also mention that information was obtained from a series of pressure holding tests using the pressuremeter and that plots of pressuremeter moduli vs. duration of pressure applied are available upon request. Also the CPT data is available on a floppy disk.

Sincerely,

Robert Gibbens

Jean-Louis Briaud, Chairman, Settlement '94

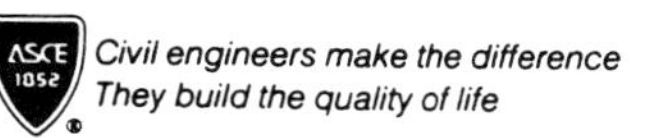

345 East 47th Street
New York, NY 10017
(212) 705-7496

SETTLEMENT '94
GEOTECHNICAL DIVISION

SYMPOSIUM UPDATE:

Spread Footing "As-Built" Dimensions

Dear Predictor:

The following is a list of "As-Built" dimensions for the spread footing to be load tested at the National Geotechnical Experimentation Site on the Riverside Campus of Texas A&M University. The footings were poured on September 8 and 9, 1993; the load test is planned for early October. The As-Built dimensions are:

	Length x Width (m)	Thickness (m)	Embedment Depth
Footing 1:	3.004 x 3.004	1.219	0.762
Footing 2:	1.505 x 1.492	1.219	0.762
Footing 3:	3.023 x 3.016	1.346	0.889
Footing 4:	2.489 x 2.496***	1.219	0.762
Footing 5:	0.991 x 0.991	1.168	0.711

*** The design dimension of this footing was 2.0 m. The actual constructed dimension was considerably larger. Please make note of this. Also, the weather this summer at the site has been very dry, with no rain.

Sincerely,

Robert Gibbens

Jean-Louis Briaud, Chairman, Settlement '94

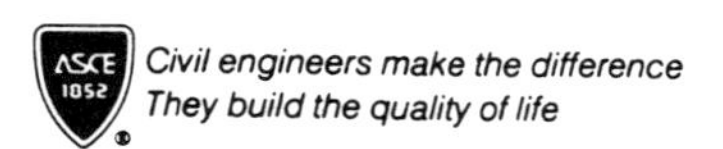

November 11, 1993

TO: Settlement 94 Prediction Event Participants

SUBJECT: Time dependent pressuremeter data.

Dear Predictor:

Attached is the data which gives the pressuremeter modulus as a function of time. This data was obtained by holding the pressure for 10 minutes during a test. The pressure was close to the end of the linear part of the PMT curve in those preboring pressuremeter tests. The n_c values are the slopes of the regression lines on those log-log plots.

You may find this data useful for the time dependent predictions. Please remember the deadline of December 1, 1993. We will see you in June 1994.

Best wishes,

Jean-Louis Briaud
Chairman, Settlement 94

CIVIL College Station, Texas 77843-3136 • (409) 845-4414; FAX (409) 845-6554

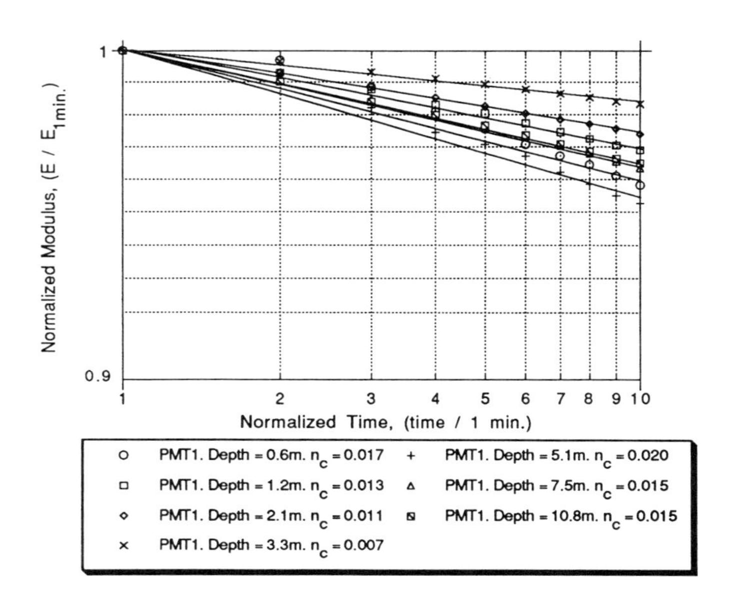

FIG. 5.13. Determination of the Creep Exponent for PMT #1

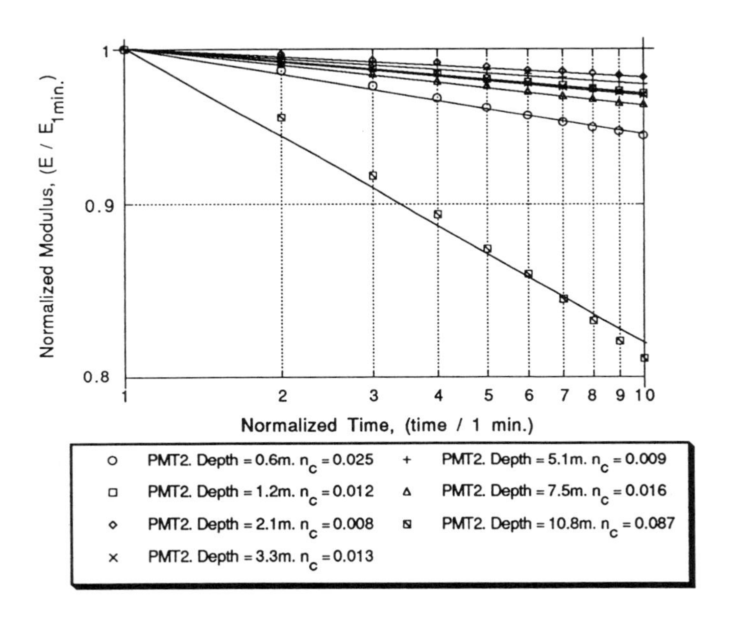

FIG. 5.14. Determination of the Creep Exponent for PMT #2

TEST AND PREDICTION RESULTS FOR FIVE LARGE SPREAD FOOTINGS ON SAND

Jean-Louis Briaud[1], Robert Gibbens[2]

ABSTRACT: Five square spread footings were load tested to 0.15 m of penetration. The footings ranged in size from 1 m to 3 m, were embedded 0.75 m, and were loaded vertically at their center. A detailed soil investigation program was undertaken and led to numerous in situ and laboratory soil test results. The soil test results and the spread footing details were sent to engineers around the world. Thirty-one class A predictions were received from eight different countries. The test results are shown and analyzed. The predictions are compared to the measurements and conclusions are drawn with respect to the current accuracy with which the profession can predict the behavior of footings on sand.

INTRODUCTION

Five square spread footings ranging in size from 1 m to 3 m were load tested to 0.15 m of displacement. In parallel, a prediction event was organized. The prediction event was advertized in *ASCE News* and at various conferences in early 1993. A flier was distributed worldwide to the approximately 6,000 members of the ASCE Geotechnical Division. About 150 requests for a prediction package were received. The 150 prediction packages were sent in July 1993. During the following two months, three addendums were sent: an errata on the prediction package, an as-built set of dimensions for the footings, and the time dependent modulus data from PMT tests. The prediction package and the addendums are given in this volume as separate papers. A total of 31 predictions was received from Israel, Australia, Japan, Canada, USA, Hongkong, Brazil, France, and Italy. The 31 papers follow this summary paper.

---

[1]Buchanan Prof. of Civil Engrg., Texas A&M University, College Station, TX, 77843-3136.

[2]Engineer, Geotest Engineering, Inc., 5600 Bintlif Drive, Houston, TX 77036, USA.

SOIL, FOOTINGS, AND LOAD TEST DETAILS

The soil is a medium dense fine silty silica sand. This middle Eocene sand was formed in a coastal plain environment. The grain size distribution curve is relatively uniform with most of the grain sizes between 0.5 mm and 0.05 mm. The sand is probably lightly overconsolidation by dessication of the fines and removal of about 1 m of overburden at the location of the spread footing tests. A series of soil tests was performed in the spring of 1993: classification tests, water content, unit weight, relative density, triaxial tests, resonant column tests, standard penetration tests with energy measurement, piezo cone penetrometer tests, pressuremeter tests, cross-hole wave tests, dilatometer tests, borehole shear tests, and step blade tests. The boring locations are on Fig. 3 of the prediction package. The data are described in detail in the prediction package.

The as-built dimensions of the footings are shown in Table 1. The general load test procedure (Fig. 1) consisted of applying the load in increments and of reading the footing settlement as a function of time for each load. Each load step lasted 30 minutes with settlement readings at 1, 3, 5, 7, 10, 15, 20, 25, and 30 minutes. The settlement was obtained from LVDTs located at each corner of the footings and attached to a settlement beam. The end supports of the settlement beam were resting on the ground surface and were located as listed in Table 2. Therefore, the measured settlement was the difference between the average of the movements of the footing corners and the movement of the settlement beam supports. The load was measured with a calibrated load cell. The load test system had a working load capacity of 12,000 kN. During the load tests, inclinometer measurements were taken in inclinometer casings surrounding the spread footings. Displacements at depths of 0.5B, 1B, 2B below the footing level were also recorded by using a telltale system. The reaction shafts were 0.91 m in diameter with a 2.7 m diameter bell in the shale (Fig. 1).

Table 1. Footing Dimensions

Footing no.	Length x Width (m)	Thickness (m)	Embedment Depth (m)	In Text, Referred to as
1	3.004 x 3.004	1.219	0.762	3 m North footing
2	1.505 x 1.492	1.219	0.762	1.5 m footing
3	3.023 x 3.016	1.346	0.889	3 m South footing
4	2.489 x 2.496	1.219	0.762	2.5 m footing
5	0.991 x 0.991	1.168	0.711	1 m footing

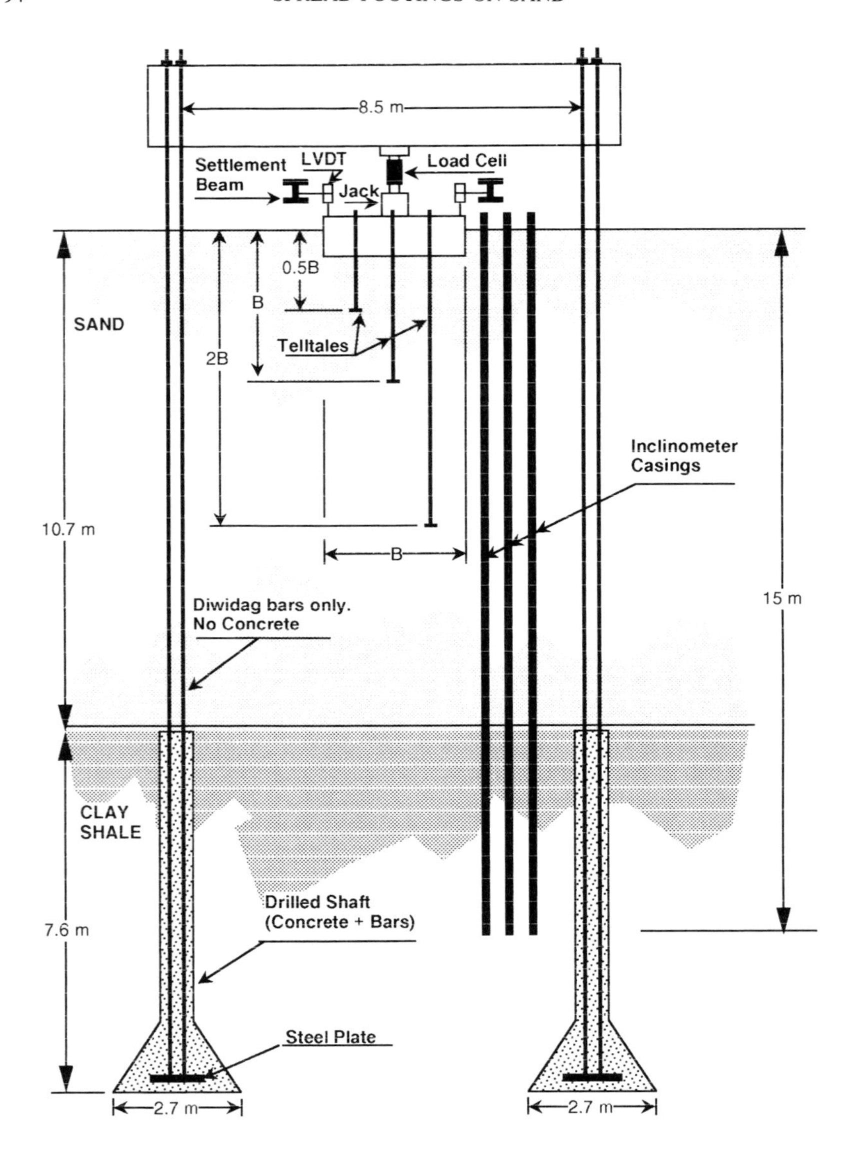

FIG. 1. Load Test Setup.

Table 2. Distance to Settlement Beam Supports

Footings	Distance Between Settlement Beam Supports
1 m	7.63 m (7.63 B)
1.5 m	9.30 m (6.20 B)
2.5 m	10.43 m (4.17 B)
3 m (South)	13.98 m (4.66 B)
3 m (North)	13.98 m (4.66 B)

LOAD TEST RESULTS

The load settlement curves for the five footings were generated by plotting all the measurements taken (total history). These included the various settlement readings as a function of time for each 30 minute load step, the unload-reload cycles performed in order to shim the jack, and the 24 hour load hold performed at working loads. Fig. 2 is an example of such a curve for the 3 m North footing. The envelope of the 30 minute readings was used to generate the best estimate of the 30 minute monotonic load-settlement curve (Figs. 3 to 7). The loads at 25 mm settlement and at 150 mm settlement were read on those curves and represented the measured values for the questions asked to the prediction event participants. The creep settlement between the one minute reading and the 30 minute reading for the two 3 m footings were obtained as the difference between the 1 minute and the 30 minute monotonic load settlement curve envelopes for those two footings. The settlement of the two 3 m footings by the year 2014 will become available in the year 2014 at which time it is suggested that another conference be held. The measured answers to the questions asked are presented in Table 3.

PREDICTION RESULTS AND COMPARISONS

A total of 31 written responses were obtained: 16 from academics and 15 from consultants (Fig. 8). A summary of the methods used is presented in Fig. 9 and of the soil tests used in Fig. 10. Table 4 and 5 give all the predictions as read by the authors in the papers sent by the predictors.

The predicted results were compared to the measured results by presenting a series of frequency distribution plots (Figs. 11 to 24). Figs. 11 through 15 give the frequency of occurrence of the ratio of the predicted load at 25 mm of settlement over the measured load at 25 mm of settlement for the five footings. Figs. 16 through 20 give the frequency of occurrence of the ratio of the predicted load at 150 mm of settlement over the measured load at 150 mm of settlement for the five footings. Table 6 gives the ratios of $Q_{predicted} / Q_{measured}$ for 25 mm of settlement while Table 7 gives the same ratio for 150 mm of settlement.

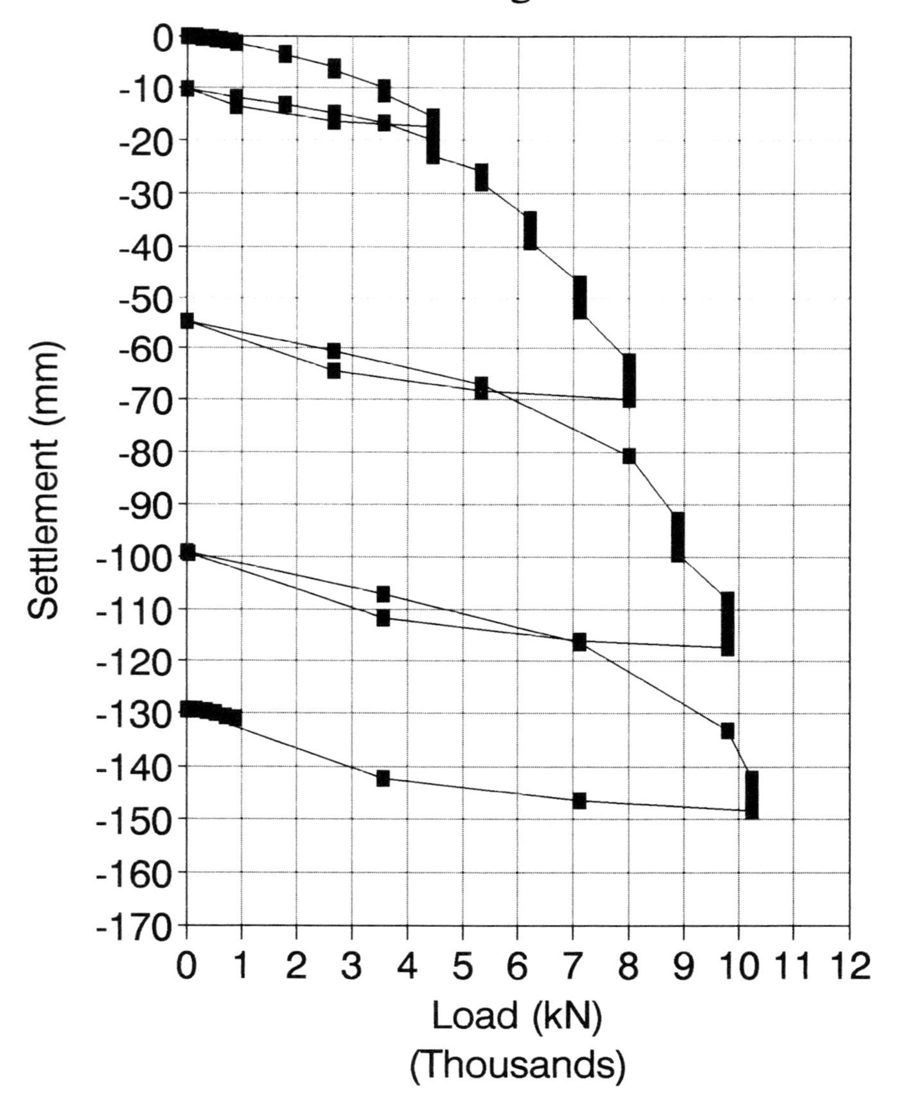

FIG. 2. Load Settlement Curve for 3 m North Footing: Total History.

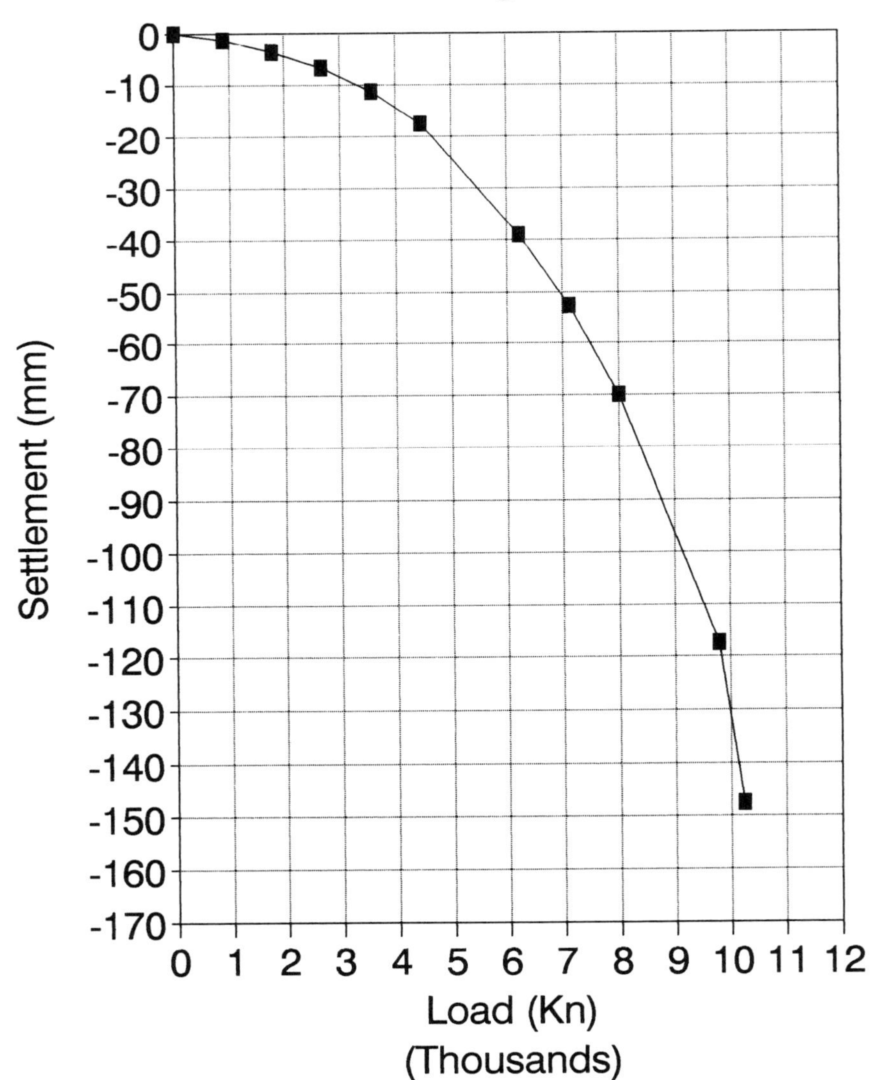

FIG. 3. Load Settlement Curve for 3 m North Footing: 30 Minute Envelope.

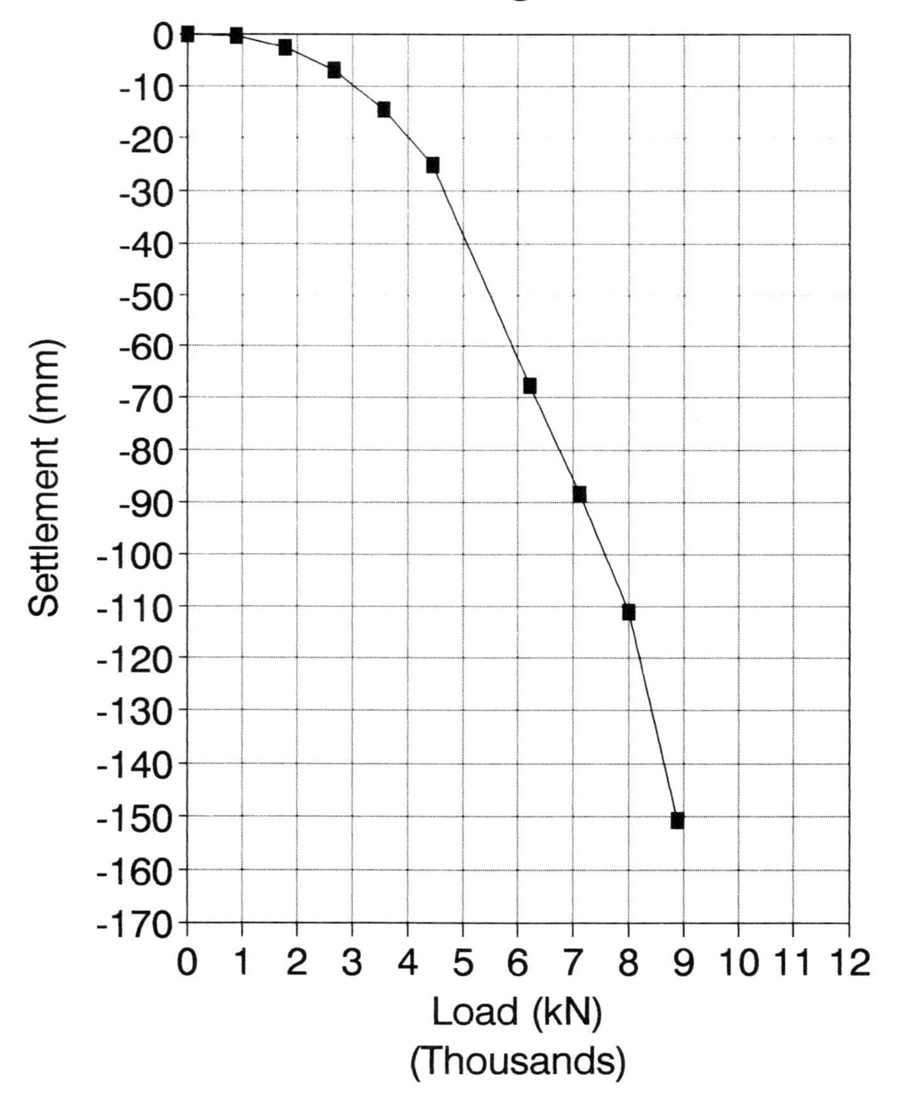

FIG. 4. Load Settlement Curve for 3 m South Footing: 30 Minute Envelope.

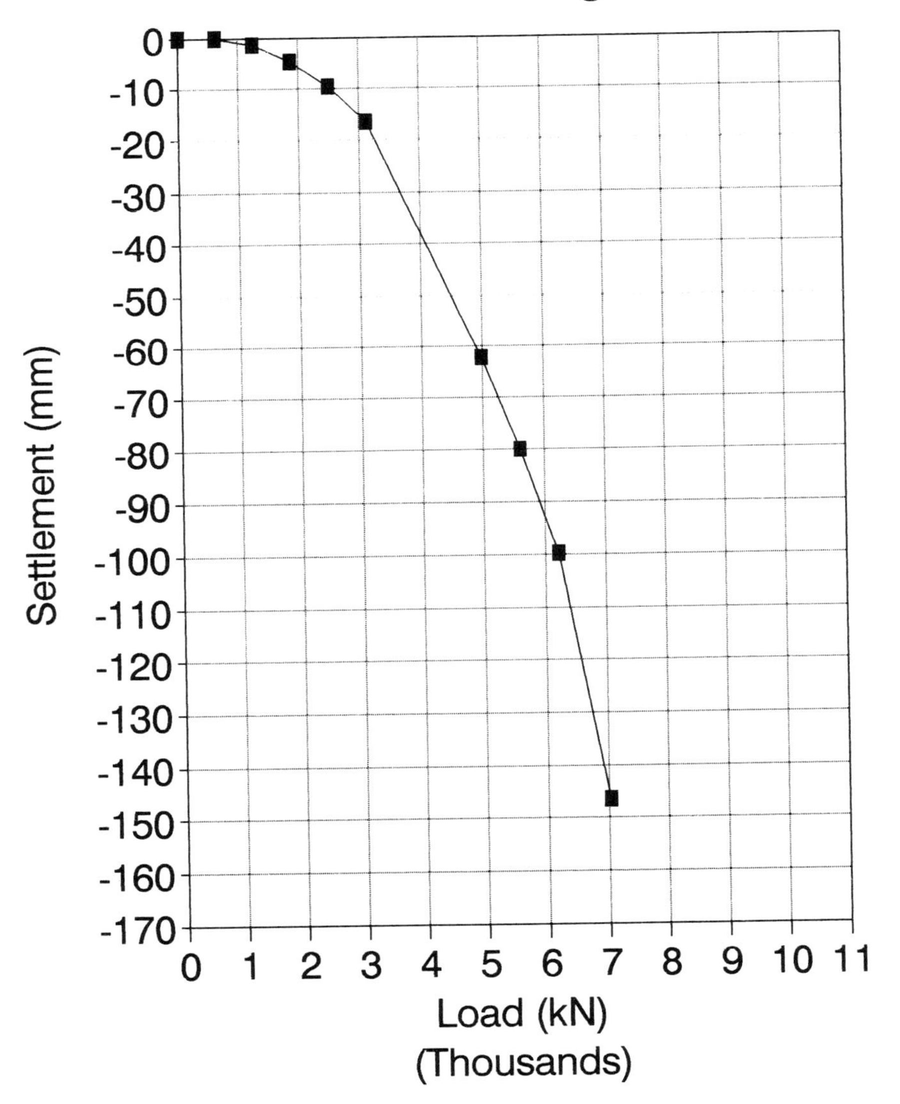

FIG. 5. Load Settlement Curve for 2.5 m Footing: 30 Minute Envelope.

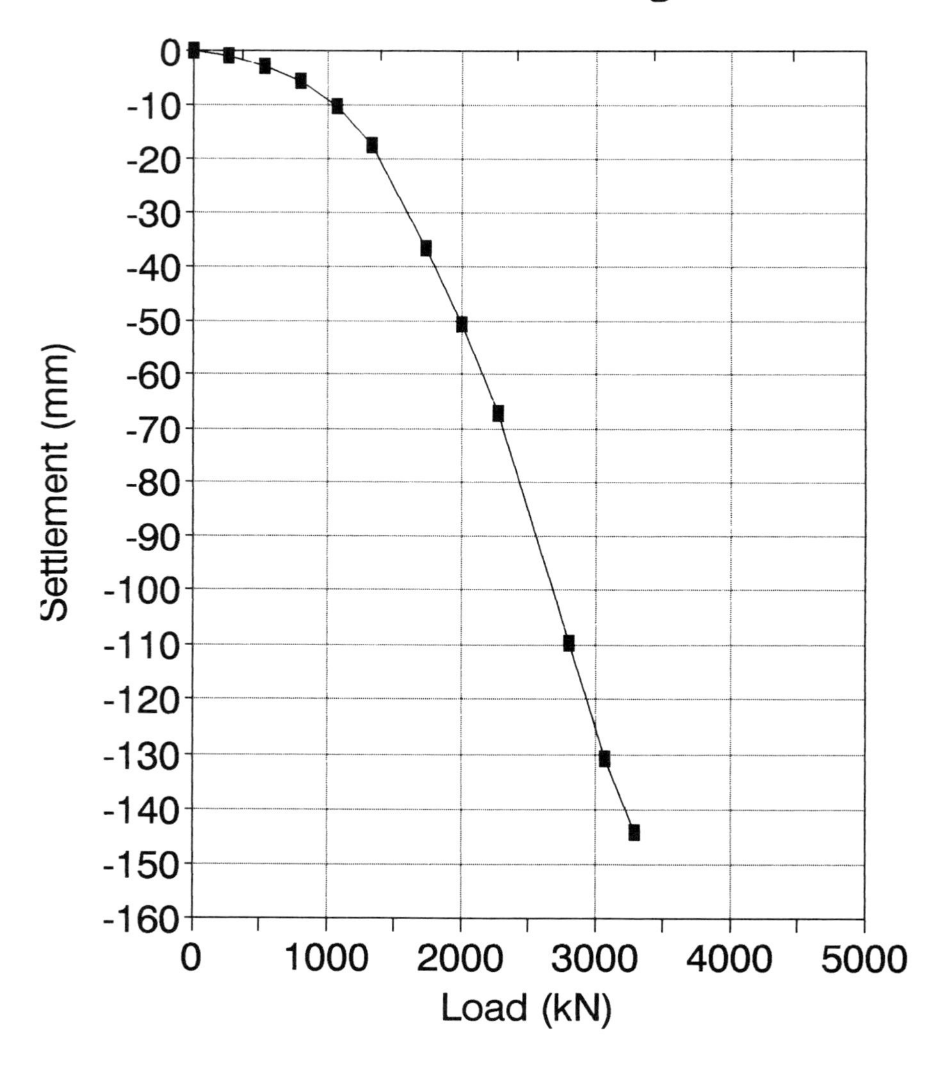

FIG. 6. Load Settlement Curve for 1.5 m Footing: 30 Minute Envelope.

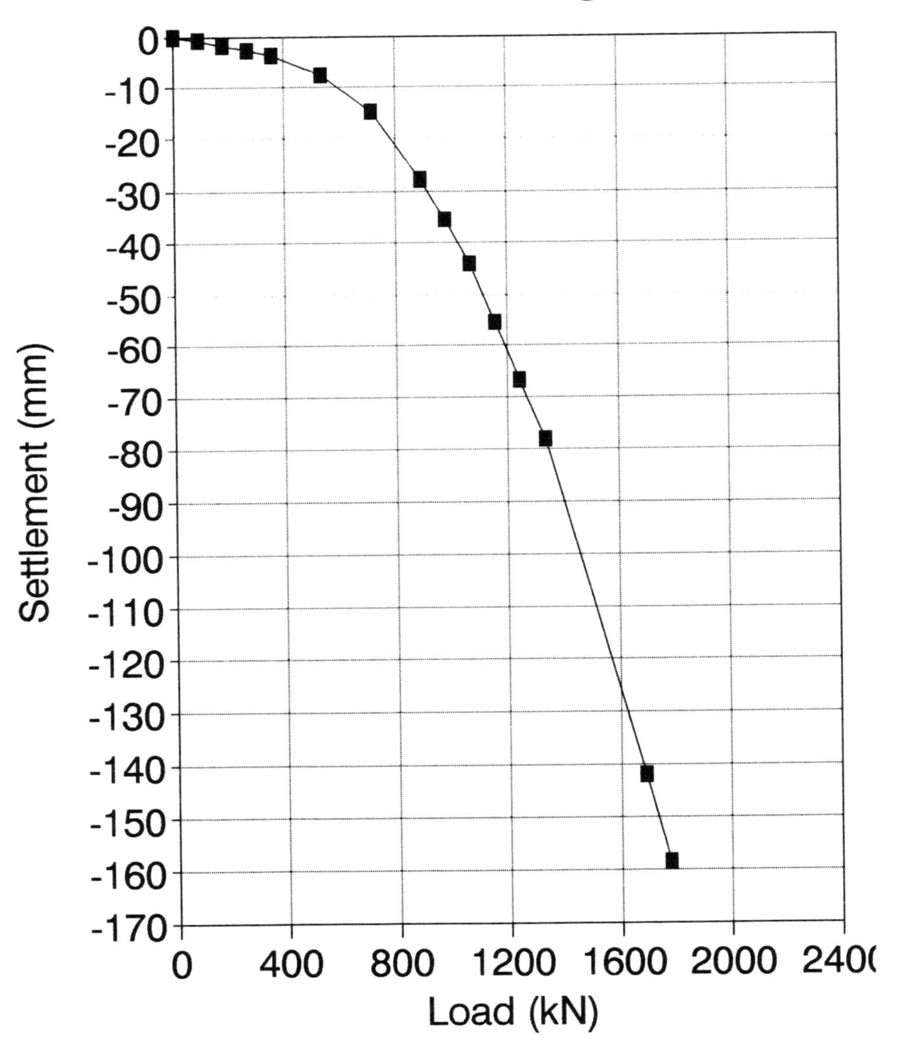

FIG. 7. Load Settlement Curve for 1 m Footing: 30 Minute Envelope.

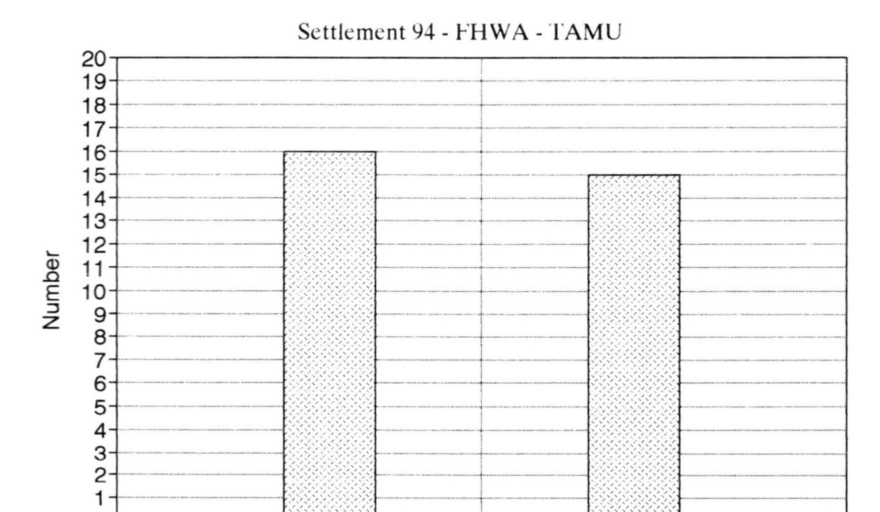

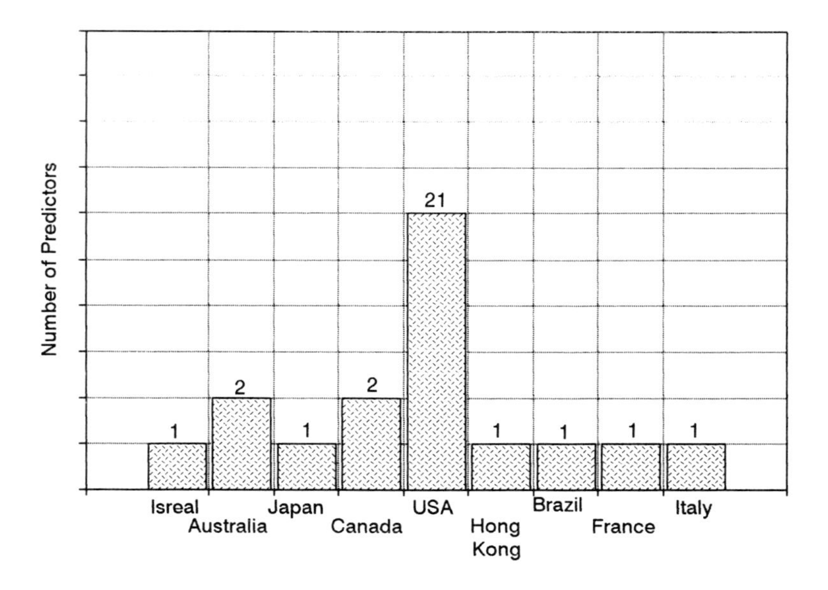

FIG. 8. Participants.

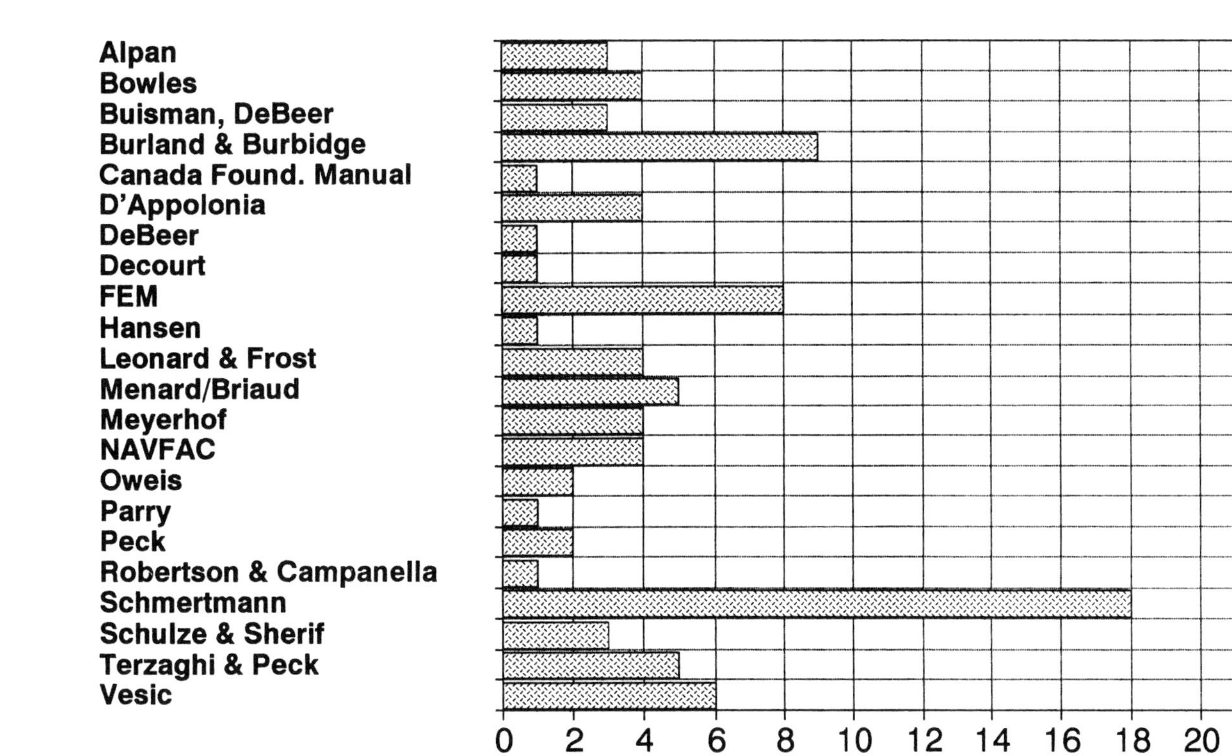

FIG. 9. Methods Used.

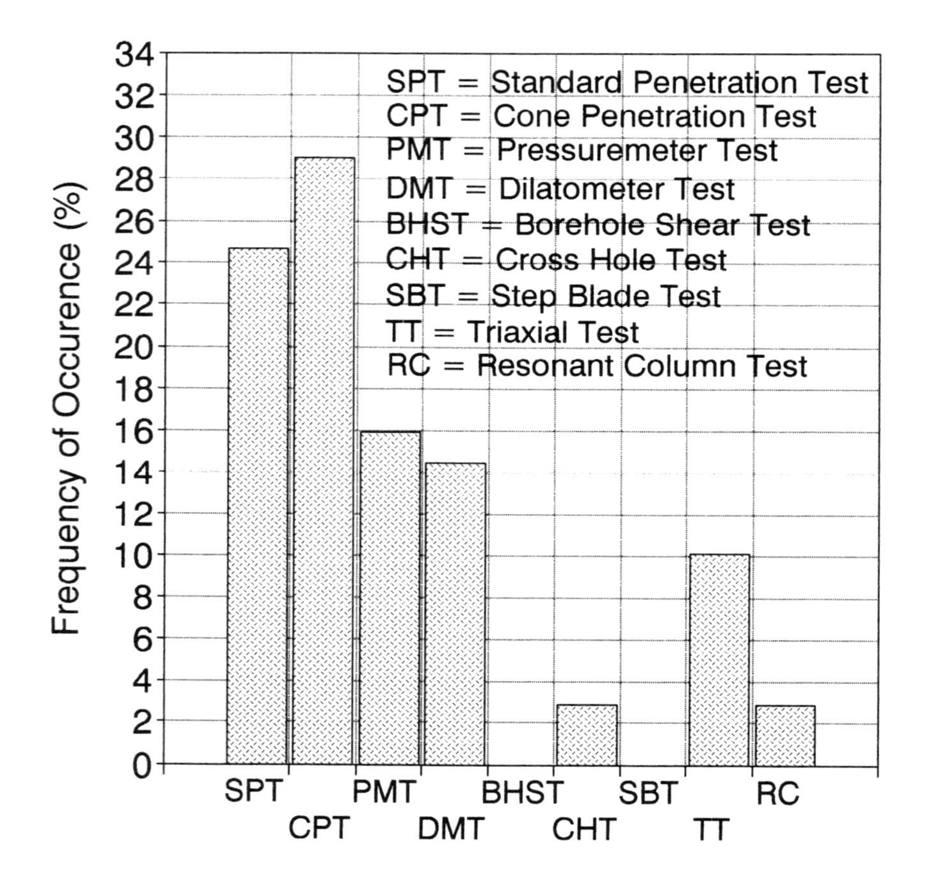

FIG. 10. Soil Tests Used.

Table 4. Prediction Results for Q_{25} and Q_{150}.

No.	Authors	Q(25) 1m kN	Q(25) 1.5m kN	Q(25) 2.5m kN	Q(25) 3.0m(s) kN	Q(25) 3.0m(n) kN	Q(150) 1m kN	Q(150) 1.5m kN	Q(150) 2.5m kN	Q(150) 3.0m(s) kN	Q(150) 3.0m(n) kN	No. of Predictions within 20% of meas. (max = 10)
1	Wiseman	670	1300	3000	3900	3900	1120	2600	7900	11300	11300	5
2	Poulos	800	5040	2560	2790	3690	1120	11340	9680	14580	12690	1
3	Siddiquee	59	116	295	407	415	200	422	1086	1326	1502	0
4	Silvestri	448	771	1488	1929	1929	1085	2143	5060	6857	6857	0
5	Horvath	900	1450	3125	4500	4500	1650	4200	13500	23000	20000	6
6	Thomas	374	450	1786	1226	2835	457	650	2927	1633	4847	0
7	Surendra	800	2590	4020	4780	4780	800	2640	4690	13960	13960	4
8	Chang	150	320	1000	1400	2370	450	500	2600	3600	5000	0
9	Brahma	720	775	1850	1550	3025	1100	1990	9630	8450	15200	2
10	Floess	700	1500	4300	5600	6400	1000	2400	7700	11500	11600	5
11	Boone	600	1100	2400	1750	3000	760	2150	7000	10000	10000	3
12	Cooksey	550	900	2100	3750	3700	900	2000	6400	9000	9000	4
13	Scott	620	1250	2380	3000	3300	850	2400	9600	15800	15800	1
14	Townsend	404	2100	3500	5400	6080	1603	7195	12804	16620	22835	4
15	Foshee	423	564	1190	655	1838	2104	2720	5949	2641	9016	4
16	Mesri	925	1475	2775	3325	3325	-	-	-	-	-	3 of 5
17	Ariemma	1100	1165	3750	1250	5480	3965	5155	14980	5220	21575	2
18	Tand	850	1550	3000	3700	3700	960	2170	6020	8830	8740	7
19	Funegard	850	1570	3370	4470	4470	960	2170	6020	8830	8740	8
20	Deschamps	900	1800	2700	5200	5200	1500	3400	5400	10800	10800	8
21	Altaee	600	1100	1900	2300	2300	2500	4300	8000	10000	10000	3
22	Decourt	779	1295	2740	4658	4290	2360	4490	16440	27945	25740	4
23	Mayne	330	950	3350	5200	5200	395	1260	5150	8450	8450	5
24	Kuo	320	650	2000	2650	3100	420	970	3500	5400	5500	0
25	Shahrour	275	450	740	1530	1530	887	2397	3720	7560	7560	1
26	Abid	550	1000	2600	3600	3300	1600	3000	8000	12000	11500	5
27	Utah State	838	1644	3087	4808	4668	900	2119	7375	15143	13943	6
28	Gottardi	935	2008	4271	5526	5446	1093	3143	10918	17379	16587	4
29	Chua	313	540	1009	1452	1456	501	937	2413	3320	3345	0
30	Bhowmik	550	800	2000	2800	3000	1100	2500	9000	14000	15000	0
31	Diyaljee	422	788	1801	2552	2526	613	1576	5280	7293	7219	1
Mean		605	1258	2454	3150	3573	1165	2831	7291	10415	11477	5
Standard Deviation		257	899	1028	1582	1439	771	2163	3743	6063	5799	
Measured Value		850	1500	3600	4500	5200	1740	3400	7100	9000	10250	

Table 5. Prediction Results for Δs and S_{2014}.

No.	Authors	Delta S(s) 1-30 mm	Delta S(n) 1-30 mm	Settlement (s) 2014 mm	Settlement (n) 2014 mm
1	Wiseman	13	13	36	36
2	Poulos	1	1	37	37
3	Siddiquee	-	-	-	-
4	Silvestri	1.3	1.3	31	31
5	Horvath	1	1	10	10
6	Thomas	1.5	1.5	35	35
7	Surendra	3	3	36.5	36.5
8	Chang	2.5	1.5	6	3
9	Brahma	2.8	2.4	32	31.5
10	Floess	1	1	33	33
11	Boone	2	2	32	32
12	Cooksey	1.3	1.3	30	30
13	Scott	1	1	37	37
14	Townsend	-	-	-	-
15	Foshee	-	-	37	38
16	Mesri	-	-	32	32
17	Ariemma	21	20.5	102	105
18	Tand	2.8	2.8	38	38
19	Funegard	2.8	2.8	32	32
20	Deschamps	1.3	1.3	30	30
21	Altaee	3	3	-	-
22	Decourt	2	2	44.3	44.3
23	Mayne	-	-	-	-
24	Kuo	16	16	27	27
25	Shahrour	-	-	-	-
26	Abid	1.5	1.5	25	25
27	Utah State	3	3	40.58	40.54
28	Gottardi	1.1	1.1	29.7	29.7
29	Chua	5.8	5.8	187.3	186.9
30	Bhowmik	1.6	1.4	35	32
31	Diyaljee	6	6	64	64
Mean		4	4	42	41
Standard Deviation		5	5	34	35
Measured Value		2.9	2.4	-	-

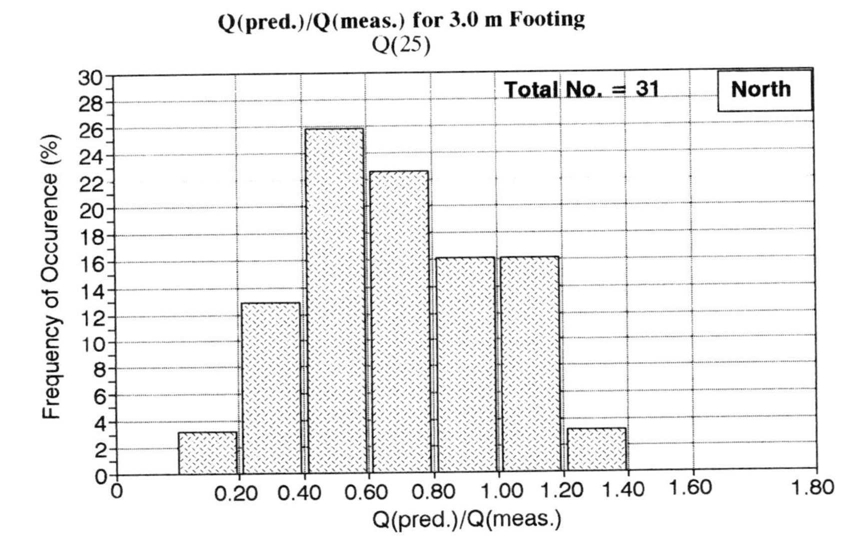

FIG. 11. Distribution of $Q_{PRED.}/Q_{MEAS.}$ for 25 mm Settlement: 3 m North Footing.

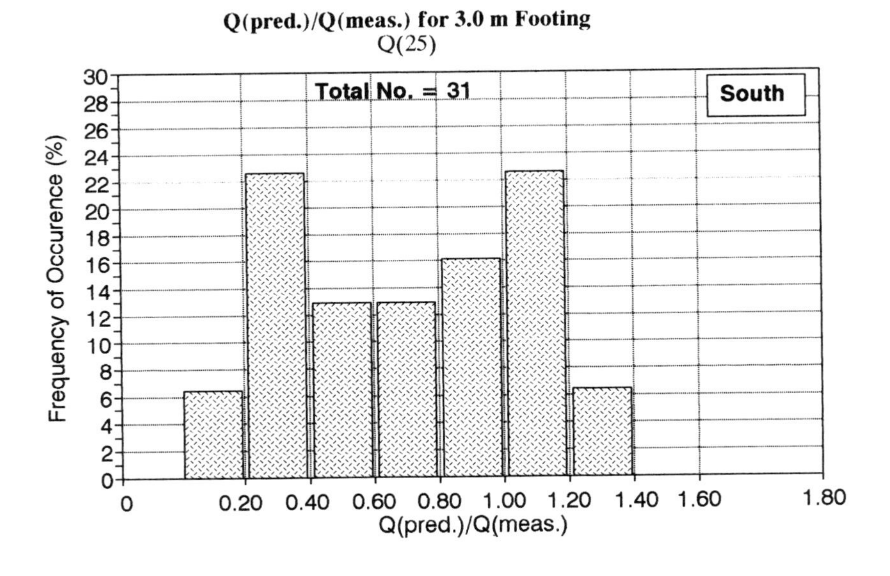

FIG. 12. Distribution of $Q_{PRED.}/Q_{MEAS.}$ for 25 mm Settlement: 3 m South Footing.

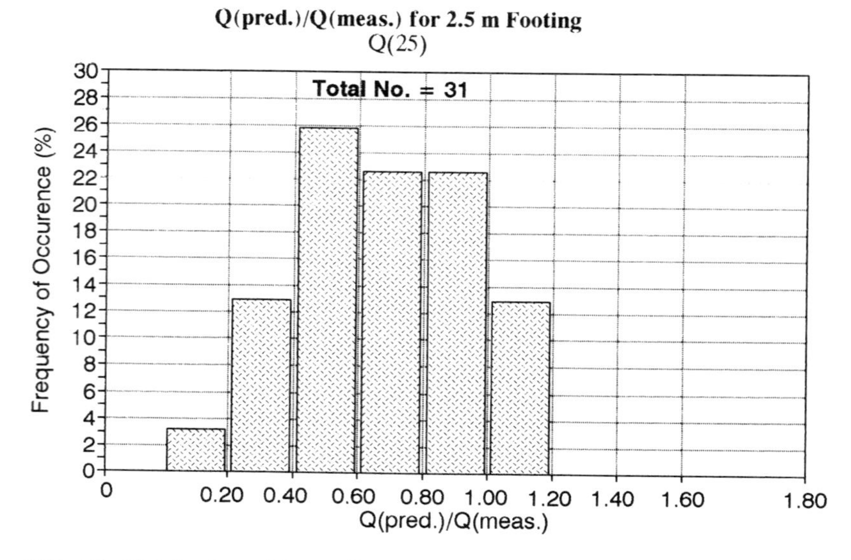

FIG. 13. Distribution of $Q_{PRED.}/Q_{MEAS.}$ for 25 mm Settlement: 2.5 m Footing.

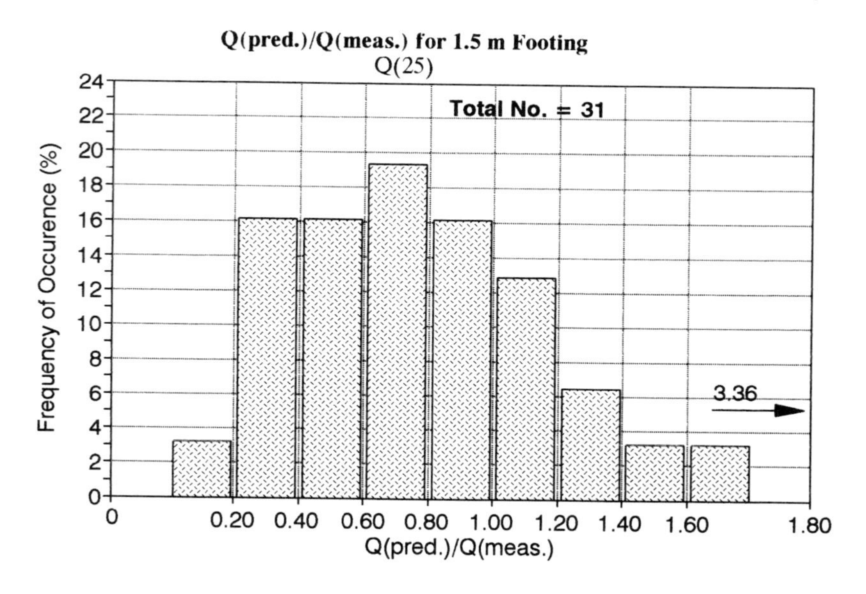

FIG. 14. Distribution of $Q_{PRED.}/Q_{MEAS.}$ for 25 mm Settlement: 1.5 m Footing.

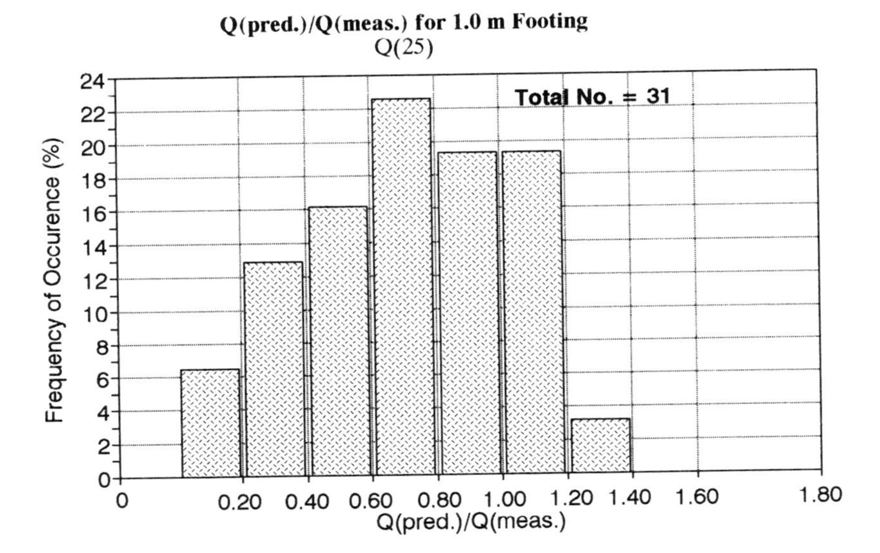

FIG. 15. Distribution of $Q_{PRED.}/Q_{MEAS.}$ for 25 mm Settlement: 1 m Footing.

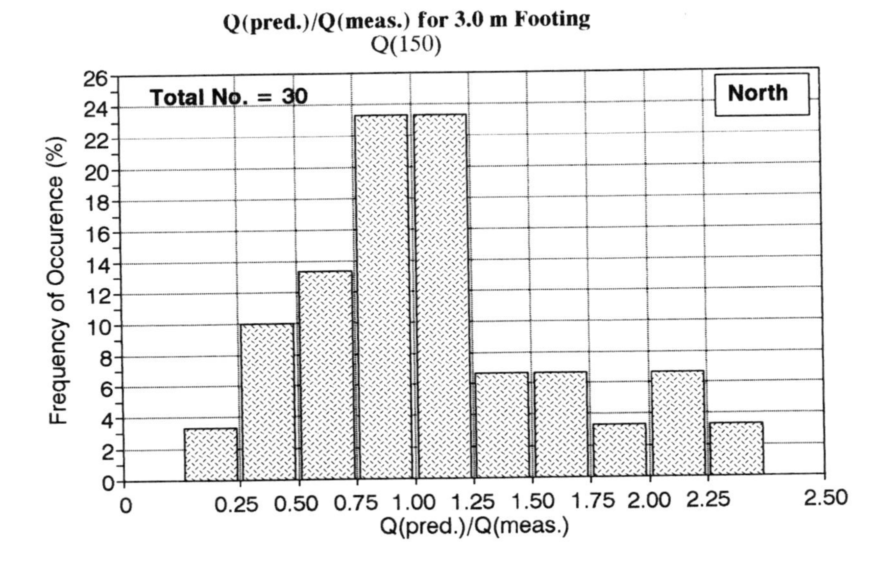

FIG. 16. Distribution of $Q_{PRED.}/Q_{MEAS.}$ for 150 mm Settlement: 3 m North Footing.

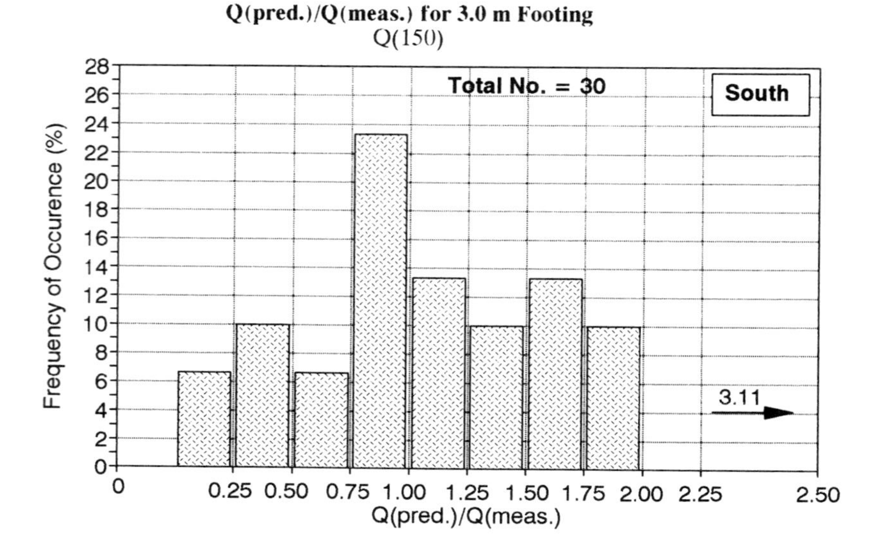

FIG. 17. Distribution of $Q_{PRED.}/Q_{MEAS.}$ for 150 mm Settlement: 3 m South Footing.

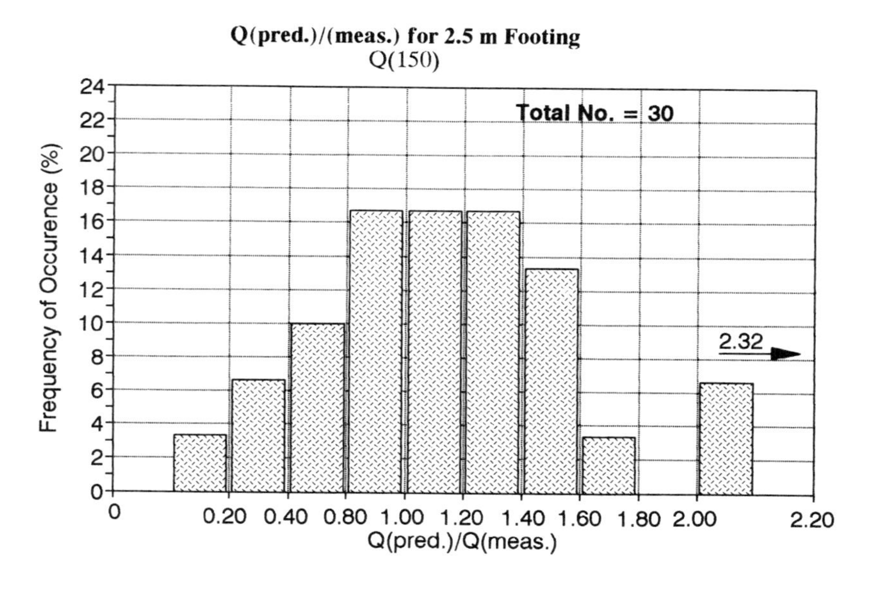

FIG. 18. Distribution of $Q_{PRED.}/Q_{MEAS.}$ for 150 mm Settlement: 2.5 m Footing.

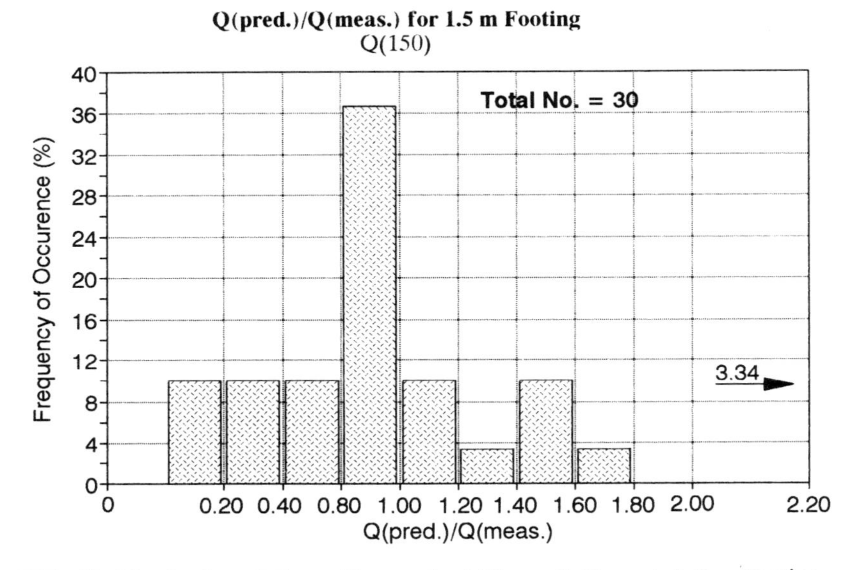

FIG. 19. Distribution of $Q_{PRED.}/Q_{MEAS.}$ for 150 mm Settlement: 1.5 m Footing.

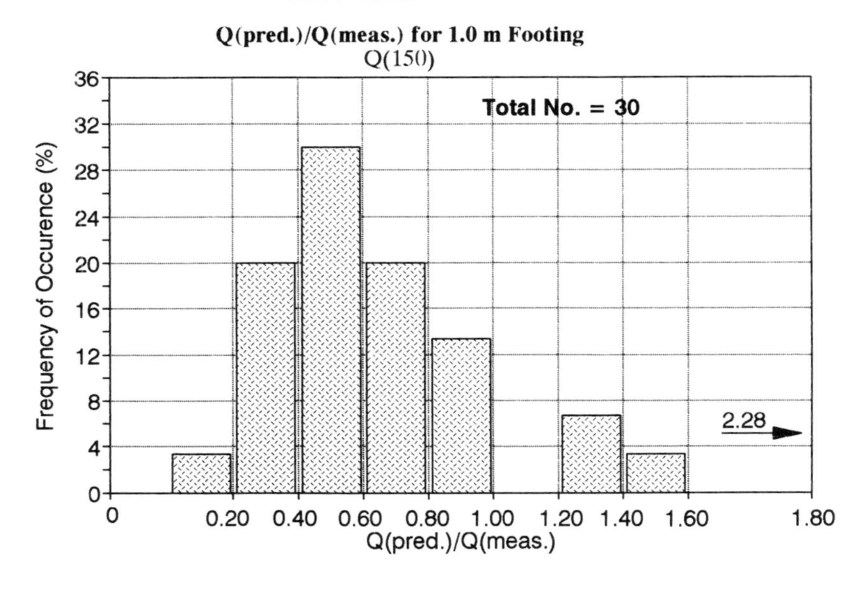

FIG. 20. Distribution of $Q_{PRED.}/Q_{MEAS.}$ for 150 mm Settlement: 1 m Footing.

Table 6. Ratio of Q_{PRED}/Q_{MEAS} for 25 mm of Settlement.

No	Authors	Qpred/Qmeas 25mm 1.0 m	Qpred/Qmeas 25mm 1.5m	Qpred/Qmeas 25mm 2.5m	Qpred/Qmeas 25mm 3.0m(s)	Qpred/Qmeas 25mm 3.0m(n)
1	Wiseman	0.788	0.867	0.833	0.867	0.750
2	Poulos	0.941	3.360	0.711	0.620	0.710
3	Siddiquee	0.069	0.077	0.082	0.090	0.080
4	Silvestri	0.527	0.514	0.413	0.429	0.371
5	Horvath	1.059	0.967	0.868	1.000	0.865
6	Thomas	0.440	0.300	0.496	0.272	0.545
7	Surendra	0.941	1.727	1.117	1.062	0.919
8	Chang	0.176	0.213	0.278	0.311	0.456
9	Brahma	0.847	0.517	0.514	0.344	0.582
10	Floess	0.824	1.000	1.194	1.244	1.231
11	Boone	0.706	0.733	0.667	0.389	0.577
12	Cooksey	0.647	0.600	0.583	0.833	0.712
13	Scott	0.729	0.833	0.661	0.667	0.635
14	Townsend	0.475	1.400	0.972	1.200	1.169
15	Foshee	0.498	0.376	0.331	0.146	0.353
16	Mesri	1.088	0.983	0.771	0.739	0.639
17	Ariemma	1.294	0.777	1.042	0.278	1.054
18	Tand	1.000	1.033	0.833	0.822	0.712
19	Funegard	1.000	1.047	0.936	0.993	0.860
20	Deschamps	1.059	1.200	0.750	1.156	1.000
21	Altaee	0.706	0.733	0.528	0.511	0.442
22	Decourt	0.916	0.863	0.761	1.035	0.825
23	Mayne	0.388	0.633	0.931	1.156	1.000
24	Kuo	0.376	0.433	0.556	0.589	0.596
25	Shahrour	0.324	0.300	0.206	0.340	0.294
26	Abid	0.647	0.667	0.722	0.800	0.635
27	Utah State	0.986	1.096	0.858	1.068	0.898
28	Gottardi	1.100	1.339	1.186	1.228	1.047
29	Chua	0.368	0.360	0.280	0.323	0.280
30	Bhowmik	0.647	0.533	0.556	0.622	0.577
31	Diyaljee	0.496	0.525	0.500	0.567	0.486
Mean		0.71	0.84	0.68	0.70	0.69
Standard Deviation		0.30	0.60	0.29	0.35	0.28
Measured Value		1	1	1	1	1

Table 7. Ratio of Q_{PRED}/Q_{MEAS} for 150 mm of Settlement.

No	Authors	Qpred/Qmeas 150mm 1.0 m	Qpred/Qmeas 150mm 1.5m	Qpred/Qmeas 150mm 2.5m	Qpred/Qmeas 150mm 3.0m(s)	Qpred/Qmeas 150mm 3.0m(n)
1	Wiseman	0.644	0.765	1.113	1.256	1.102
2	Poulos	0.644	3.335	1.363	1.620	1.238
3	Siddiquee	0.115	0.124	0.153	0.147	0.147
4	Silvestri	0.624	0.630	0.713	0.762	0.669
5	Horvath	0.948	1.235	1.901	2.556	1.951
6	Thomas	0.263	0.191	0.412	0.181	0.473
7	Surendra	0.460	0.776	0.661	1.551	1.362
8	Chang	0.259	0.147	0.366	0.400	0.488
9	Brahma	0.632	0.585	1.356	0.939	1.483
10	Floess	0.575	0.706	1.085	1.278	1.132
11	Boone	0.437	0.632	0.986	1.111	0.976
12	Cooksey	0.517	0.588	0.901	1.000	0.878
13	Scott	0.489	0.706	1.352	1.756	1.541
14	Townsend	0.921	2.116	1.803	1.847	2.228
15	Foshee	1.209	0.800	0.838	0.293	0.880
16	Mesri	0.000	0.000	0.000	0.000	0.000
17	Ariemma	2.279	1.516	2.110	0.580	2.105
18	Tand	0.552	0.638	0.848	0.981	0.853
19	Funegard	0.552	0.638	0.848	0.981	0.853
20	Deschamps	0.862	1.000	0.761	1.200	1.054
21	Altaee	1.437	1.265	1.127	1.111	0.976
22	Decourt	1.356	1.321	2.315	3.105	2.511
23	Mayne	0.227	0.371	0.725	0.939	0.824
24	Kuo	0.241	0.285	0.493	0.600	0.537
25	Shahrour	0.510	0.705	0.524	0.840	0.738
26	Abid	0.920	0.882	1.127	1.333	1.122
27	Utah State	0.517	0.623	1.039	1.683	1.360
28	Gottardi	0.628	0.924	1.538	1.931	1.618
29	Chua	0.288	0.276	0.340	0.369	0.326
30	Bhowmik	0.632	0.735	1.268	1.556	1.463
31	Diyaljee	0.352	0.464	0.744	0.810	0.704
Mean		0.65	0.81	0.99	1.12	1.08
Standard Deviation		0.45	0.64	0.55	0.69	0.59
Measured Value		1	1	1	1	1

Table 3. Measured Results

	Footing 1 3 m North	Footing 2 1.5 m	Footing 3 3 m South	Footing 4 2.5 m	Footing 5 1 m
1. Load for 25 mm of settlement Q_{25} on the 30 minute load settlement curve (kN)	5200	1500	4500	3600	850
2. Load for 150 mm of settlement Q_{150} on the 30 minute load settlement curve (kN)	10250	3400	9000	7100	1740
3. Creep settlement between 1 minutes and 30 minute for Q_{25}, Δs (mm)	2.4	X	2.9	X	X
4. Settlement in the year 2014 under Q_{25} (mm)	?	X	?	X	X

Inspection of the Q_{25} frequency distribution plots indicates that 80% of the time the predictions were on the safe side and that this number was relatively independent of the footing scale. This is an indication that the scale effect is properly taken into account in settlement analyses.

Inspection of the Q_{150} frequency distribution plots indicates that on the average 63% of the time the predictions were on the safe side. This number however varied significantly from 84% for the 1 m footing to 48% for the 3 m South footing and showed a clear decreasing trend with increasing size. This is an indication that the scale effect is not properly taken into account in predicting the load corresponding to a large displacement.

Predicting the load at 150 mm of displacement created a dilemma for the participants: could this be considered enough displacement to use bearing capacity equation or not? Most considered that the answer was "Yes." Others used the FEM with a nonlinear model or did a non linear extension of their settlement method.

If one considers all the answers which were within ±20% of the measured loads, it is possible to count how many such answers each participant had (table 2). This number would have a maximum of 10. The best results were from participants whose answers fell within the ±20% range 8 times out of 10. The two participants who achieved this remarkable result used the Menard/Briaud pressuremeter method for one and an average of 15 simple methods plus the FEM for the other.

An attempt was made to find out which method was the most consistently successful at predicting the loads at 25 mm and 150 mm. The top 10 answers for each of the loads to be predicted was studied. It was not possible to detect a clear best method. What became clear, however, is that very few participants actually used the exact procedure recommended by the authors of a method. Instead many participants used a given method but modified it by taking into account their own experience or by using part of another method. Several participants used an average of several methods as their answer.

Figs. 21 and 22 show the frequency of occurrence of the ratio of the predicted creep settlement between 1 and 30 minute under the 25 mm and 150 mm load, respectively, over the measured creep settlement under the same conditions. The results indicate a reasonable prediction considering the lack of directly useful data in the prediction package. Most participants used some form of Schmertmann's time factor or the time dependent pressuremeter data sent in the addendums.

Figs. 23 and 24 show the frequency of occurrence of the predicted settlement by the year 2014 under the load which created 25 mm of settlement at 30 minutes. The predictions indicate that an additional 10 mm should occur over the next 20 years. Provided funding can be secured, the plan is to place the load of 5200 kN on the 3 m North footing and 4500 kN on the 3 m South footing for the next 20 years and to organize another conference in the year 2014.

PREDICTED DESIGN LOADS AND FACTOR OF SAFETY

The participants predicted Q_{25} and Q_{150}. In a typical design, one uses a factor of safety of 3 on the ultimate load Q_u and uses as the design load the minimum of the load leading to 25 mm of settlement and one third of the ultimate load. For the purpose of this exercise the predicted design load for each participant was taken as:

$$Q_d = MIN\ (Q_{25(predicted)},\ Q_{150(predicted)}\ /\ 3) \qquad (1)$$

These values are listed in Table 8. The measured load at 150 mm of settlement was taken as the measured failure load and the ratio of Q_F/D_D was considered to be a safety factor. The values of this factor of safety are given in Table 9. The frequency distribution of this factor of safety for all footings is presented in Fig. 25 to 29. It shows that only in one instance the factor of safety Q_F/Q_D was less than one, and that the next worst case was 1.6. Therefore nearly all predictions lead to safe designs.

Finally the settlement that would take place under the design load was obtained. This was done by using the design load for each participant and reading the corresponding settlement on the measured load settlement curve. These settlement values are in Table 10 and are presented as frequency distributions for all footings on Figs. 30 to 34. As can be seen, most settlements are quite acceptable. A trend is noticed where the larger the footing the higher the settlement under the design load.

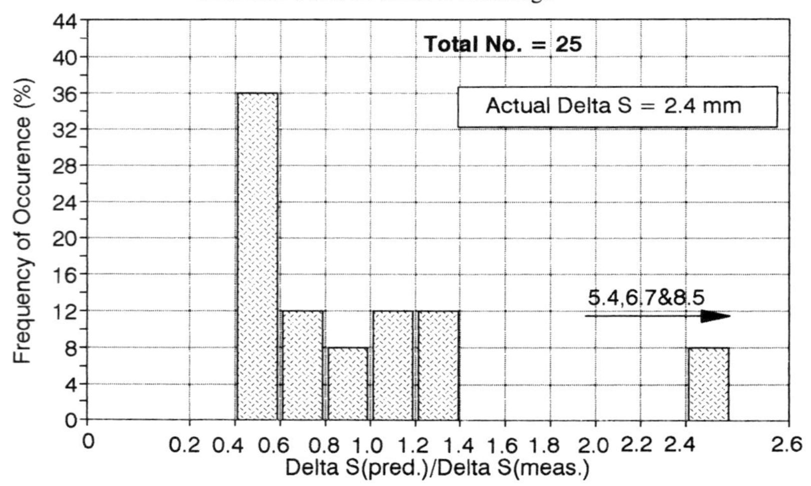

FIG. 21. Distribution of Predictions for Δs: 3 m North Footing.

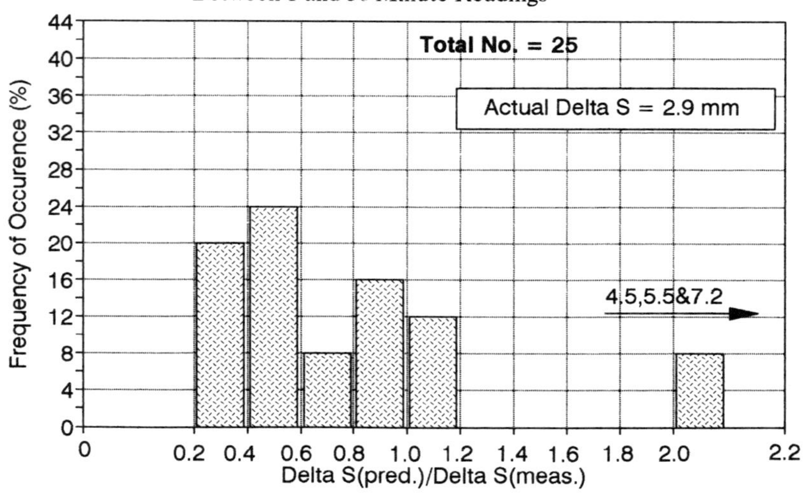

FIG. 22. Distribution of Predictions for Δs: 3 m South Footing.

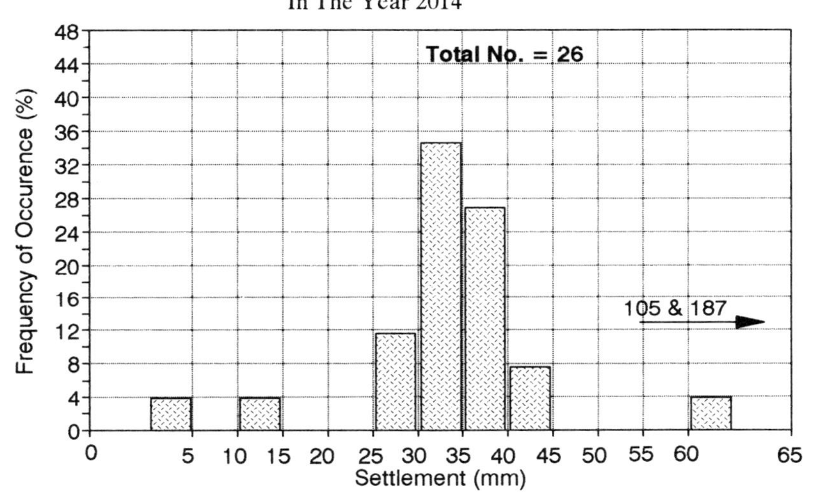

FIG. 23. Distribution of Predictions for S_{2014}: 3 m North Footing.

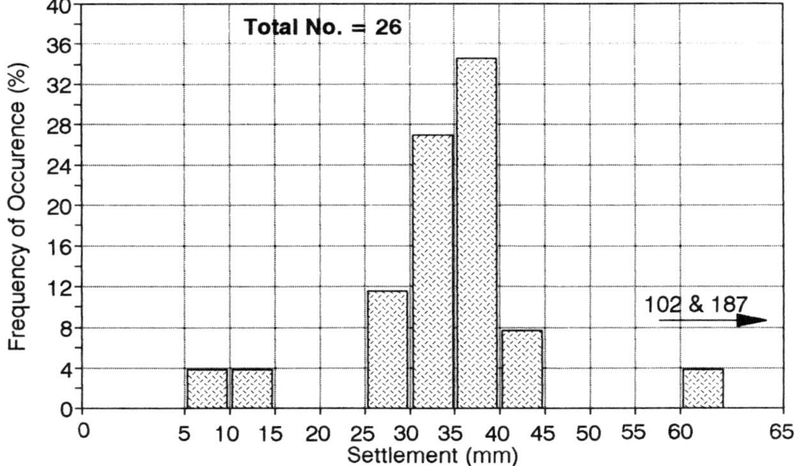

FIG. 24. Distribution of Predictions for S_{2014}: 3 m South Footing.

Table 8. Predicted Design Loads.

No.	Authors	Q(d) 1m kN	Q(d) 1.5m kN	Q(d) 2.5m kN	Q(d) 3.0m(s) kN	Q(d) 3.0m(n) kN
1	Wiseman	373	867	2633	3767	3767
2	Poulos	373	3780	2560	2790	3690
3	Siddiquee	59	116	295	407	415
4	Silvestri	362	714	1488	1929	1929
5	Horvath	550	1400	3125	4500	4500
6	Thomas	152	217	976	544	1616
7	Surendra	267	880	1563	4653	4653
8	Chang	150	167	867	1200	1667
9	Brahma	367	663	1850	1550	3025
10	Floess	333	800	2567	3833	3867
11	Boone	253	717	2333	1750	3000
12	Cooksey	300	667	2100	3000	3000
13	Scott	283	800	2380	3000	3300
14	Townsend	404	2100	3500	5400	6080
15	Foshee	423	564	1190	655	1838
16	Mesri	925	1475	2775	3325	3325
17	Ariemma	1100	1165	3750	1250	5480
18	Tand	320	723	2007	2943	2913
19	Funegard	320	723	2007	2943	2913
20	Deschamps	500	1133	1800	3600	3600
21	Altaee	600	1100	1900	2300	2300
22	Decourt	779	1295	2740	4658	4290
23	Mayne	132	420	1717	2817	2817
24	Kuo	140	323	1167	1800	1833
25	Shahrour	275	450	740	1530	1530
26	Abid	533	1000	2600	3600	3300
27	Utah State	300	706	2458	4808	4648
28	Gottardi	364	1048	3639	5526	5446
29	Chua	167	312	804	1107	1115
30	Bhowmik	367	800	2000	2800	3000
31	Diyaljee	204	525	1760	2431	2406
Mean		377	892	2042	2788	3138
Standard Deviation		229	682	866	1430	1338
Measured Value		580	1133	2367	3000	3417

Table 9. Factors of Safety $F = Q_f / Q_d$.

No.	Authors	Q(f)/Q(d) 1m	Q(f)/Q(d) 1.5m	Q(f)/Q(d) 2.5m	Q(f)/Q(d) 3.0m(s)	Q(f)/Q(d) 3.0m(n)
1	Wiseman	4.66	3.92	2.70	2.39	2.72
2	Poulos	4.66	0.90	2.77	3.23	2.78
3	Siddiquee	29.49	29.31	24.07	22.11	24.70
4	Silvestri	4.81	4.76	4.77	4.67	5.31
5	Horvath	3.16	2.43	2.27	2.00	2.28
6	Thomas	11.42	15.69	7.28	16.53	6.34
7	Surendra	6.53	3.86	4.54	1.93	2.20
8	Chang	11.60	20.40	8.19	7.50	6.15
9	Brahma	4.75	5.13	3.84	5.81	3.39
10	Floess	5.22	4.25	2.77	2.35	2.65
11	Boone	6.87	4.74	3.04	5.14	3.42
12	Cooksey	5.80	5.10	3.38	3.00	3.42
13	Scott	6.14	4.25	2.98	3.00	3.11
14	Townsend	4.31	1.62	2.03	1.67	1.69
15	Foshee	4.11	6.03	5.97	13.74	5.58
16	Mesri	1.88	2.31	2.56	2.71	3.08
17	Ariemma	1.58	2.92	1.89	7.20	1.87
18	Tand	5.44	4.70	3.54	3.06	3.52
19	Funegard	5.44	4.70	3.54	3.06	3.52
20	Deschamps	3.48	3.00	3.94	2.50	2.85
21	Altaee	2.90	3.09	3.74	3.91	4.46
22	Decourt	2.23	2.63	2.59	1.93	2.39
23	Mayne	13.22	8.10	4.14	3.20	3.64
24	Kuo	12.43	10.52	6.09	5.00	5.59
25	Shahrour	6.33	7.56	9.59	5.88	6.70
26	Abid	3.26	3.40	2.73	2.50	3.11
27	Utah State	5.80	4.81	2.89	1.87	2.21
28	Gottardi	4.78	3.25	1.95	1.63	1.88
29	Chua	10.42	10.89	8.83	8.13	9.19
30	Bhowmik	4.75	4.25	3.55	3.21	3.42
31	Diyaljee	8.52	6.47	4.03	3.70	4.26
Mean		6.64	6.29	4.72	4.99	4.43
Standard Deviation		5.23	5.90	4.12	4.63	4.13
Measured Value		3	3	3	3	3

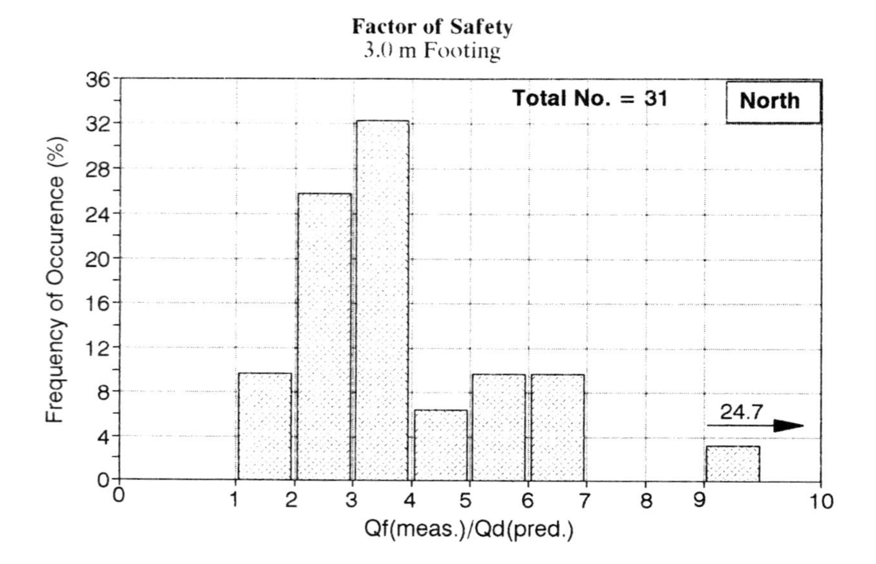

FIG. 25. Distribution of the Factor of Safety: 3 m North Footing.

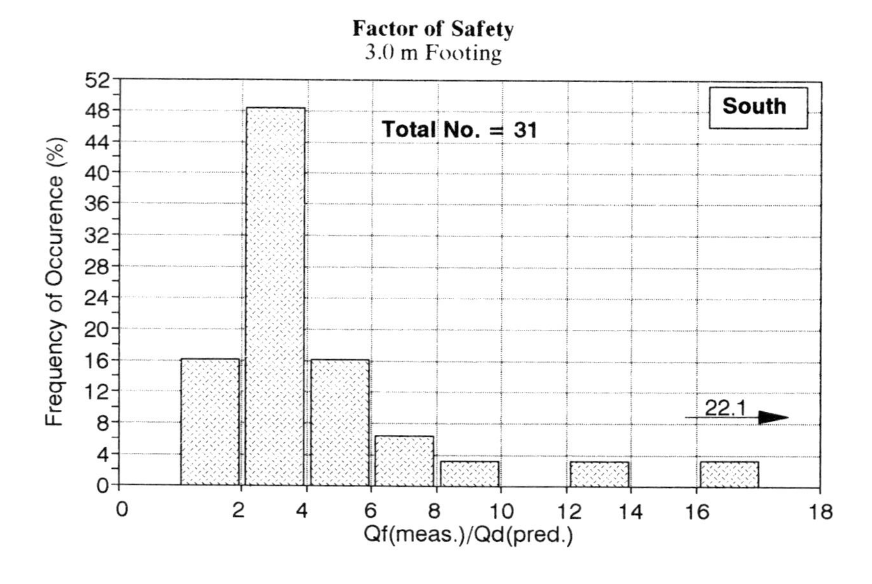

FIG. 26. Distribution of the Factor of Safety: 3 m South Footing.

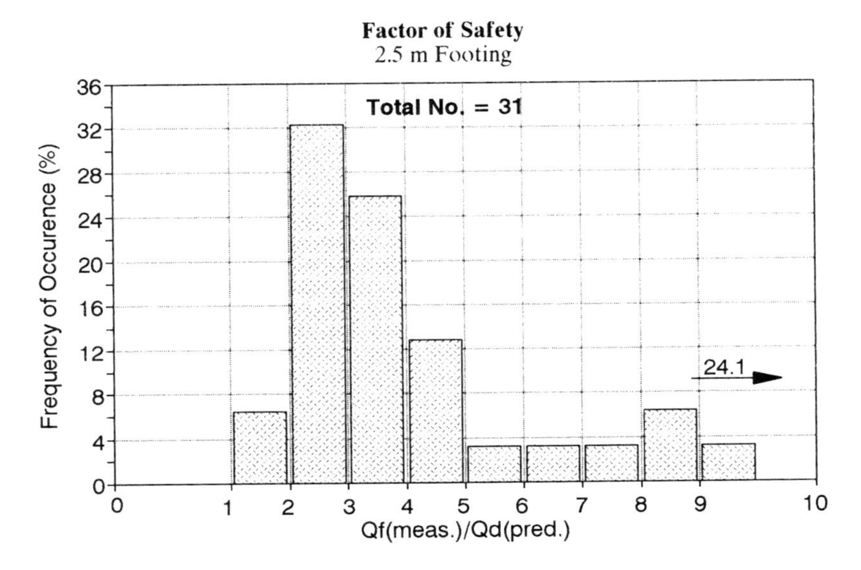

FIG. 27. Distribution of the Factor of Safety: 2.5 m Footing.

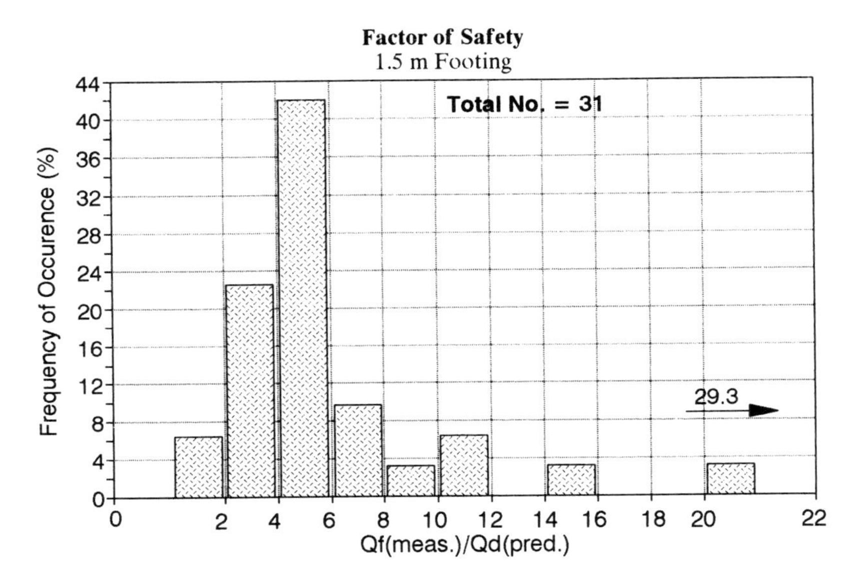

FIG. 28. Distribution of the Factor of Safety: 1.5 m Footing.

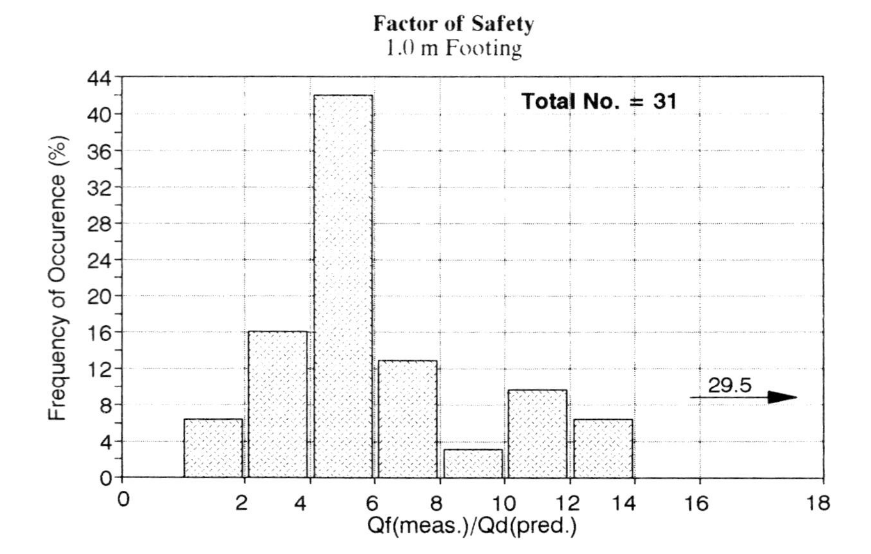

FIG. 29. Distribution of the Factor of Safety: 1 m Footing.

Table 10. Settlement under the Design Load.

No.	Authors	S(d) 1m mm	S(d) 1.5m mm	S(d) 2.5m mm	S(d) 3.0m(s) mm	S(d) 3.0m(n) mm
1	Wiseman	3	6.5	12	16	12
2	Poulos	3	170	11	8	12
3	Siddiquee	0.5	1	0	0	1
4	Silvestri	3	4	3	3	4
5	Horvath	7	21	17	25	18
6	Thomas	2	1	1	0	3
7	Surendra	2.5	7	3.5	28	19
8	Chang	2	1	1	1	20
9	Brahma	3	3.5	4.5	2	8
10	Floess	2.5	6	11	17	13
11	Boone	2.5	4.5	8	2.5	8
12	Cooksey	2.5	3.5	6	9	8
13	Scott	2.5	6	9	9	10
14	Townsend	4	55	27	47	36
15	Foshee	4.5	3	2	0	3
16	Mesri	28	24	13	12	10
17	Ariemma	48	12	33	1	30
18	Tand	2.5	4	5	9	7.5
19	Funegard	2.5	4	5	9	7.5
20	Deschamps	6	13	4	15	12
21	Altaee	10	13	4.5	5	6
22	Decourt	19	18	12.5	29	17
23	Mayne	2	2	4	8	8
24	Kuo	2	1.5	1.5	3	4
25	Shahrour	3	2	1	2	2.5
26	Abid	7.5	9	11	14	10
27	Utah State	3	4.5	9	34	20
28	Gottardi	3.5	10	29.5	50.5	29
29	Chua	1.5	1.5	0.5	1	2
30	Bhowmik	3.5	5.5	5.5	8	8.5
31	Diyaljee	2	3	4	6	6
Mean		6.1	13.5	8.4	12.1	11.5
Standard Deviation		9.5	30.9	8.4	13.4	8.6
Measured Value		9.5	12.0	8.5	10.0	10.5

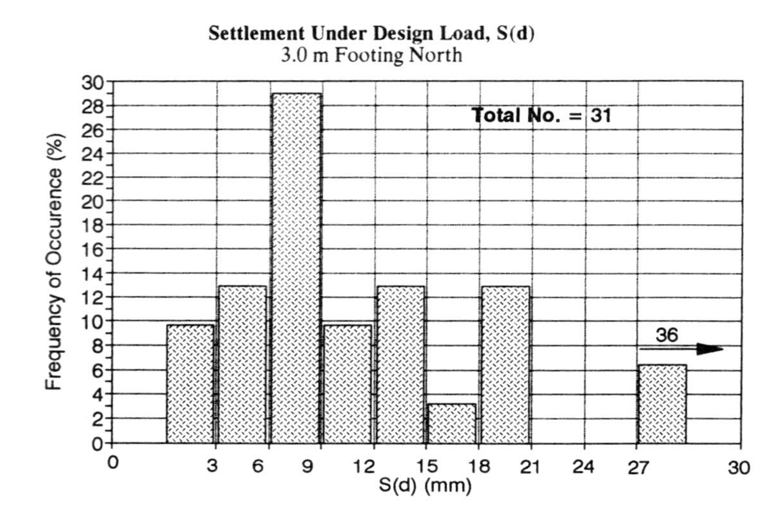

FIG. 30. Distribution of Measured Settlement under Predicted Design Load: 3 m North Footing.

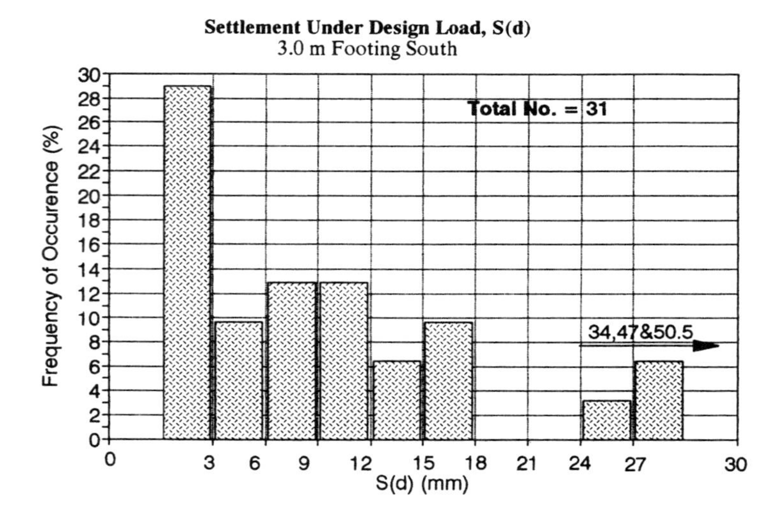

FIG. 31. Distribution of Measured Settlement under Predicted Design Load: 3 m South Footing.

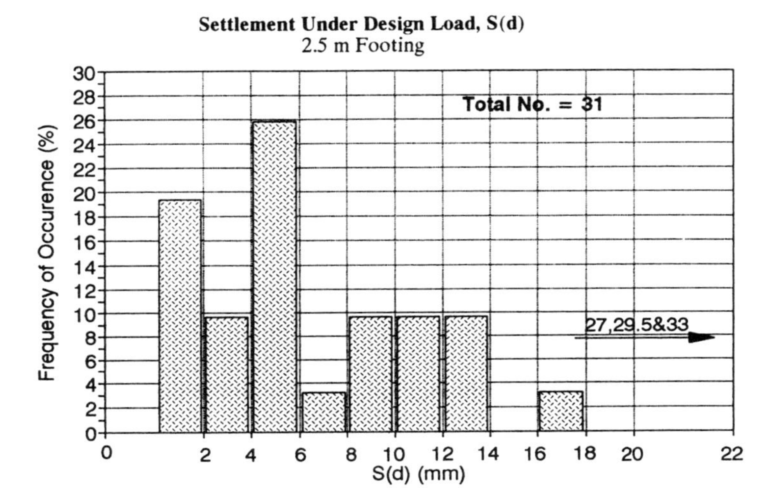

FIG. 32. Distribution of Measured Settlement under Predicted Design Load: 2.5 m Footing.

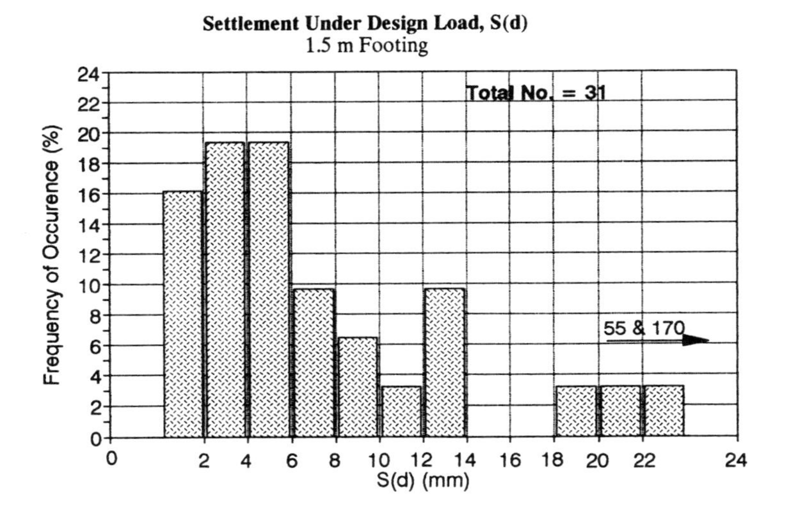

FIG. 33. Distribution of Measured Settlement under Predicted Design Load: 1.5 m Footing.

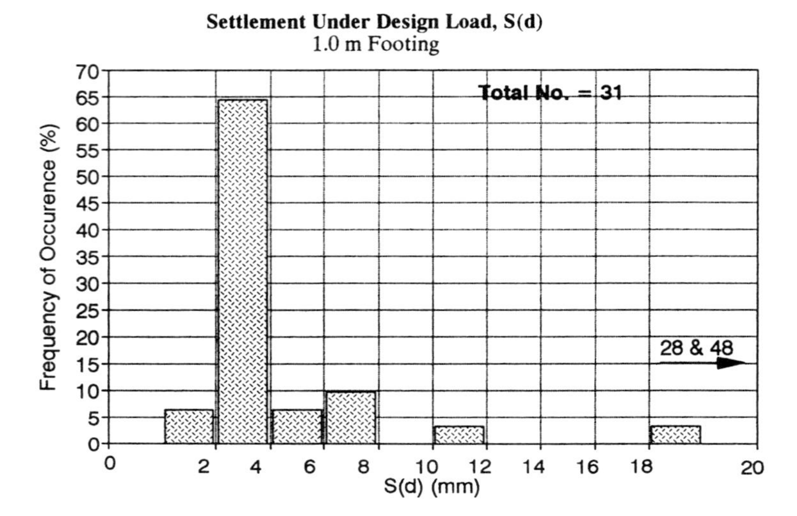

FIG. 34. Distribution of Measured Settlement under Predicted Design Load: 1 m Footing.

CONCLUSIONS

Five large spread footing load tests were performed. The square footings varied from 1 m to 3 m in width and were loaded until the settlement reached 150 mm. Participants were asked to predict the loads (Q_{25} and Q_{150}) necessary to create 25 mm and 150 mm of settlement after 30 minutes of load application as well as the creep settlement over 30 minutes for the 25 mm load Q_{25} and the settlement by the year 2014 under the 25 mm load Q_{25}. A total of 31 predictions were received from 8 different countries, half from consultants and half from academics. The conclusions reached from comparing the predictions and the measurements are as follows:

1. Nobody gave a complete set of answers which consistently fell within $\pm 20\%$ of the measured values. Two participants had 80% of their answers falling within the $\pm 20\%$ margin of error.
2. The load creating 25 mm of settlement, Q_{25}, was underestimated by 27% on the average. The predictions were 80% of the time on the safe side. The scale effect was properly predicted since this number (80%) was consistent for all sizes.
3. The load creating 150 mm of settlement, Q_{150}, was underestimated by 6% on the average. The predictions were 63% of the time on the safe side. The scale effect was not properly predicted and there was a trend towards overpredicting Q_{150} for the larger footings.
4. A large variety of methods were used and it was not possible to identify the most accurate method because most people used published methods modified by their own experience or used a combination of methods. The most popular method was Schmertmann's method using CPT data. Of all the soil tests performed, the most used one was the CPT; then came the SPT, the PMT and the DMT.
5. The creep settlement over the 30 minute load step for Q_{25} was predicted reasonably well considering the limited data available for this prediction. The average prediction for the settlement by the year 2014 under Q_{25} is 35 mm or an additional 10 mm over the next 20 years.
6. The design load Q_d for each footing and each participant was defined as MIN ($Q_{25(predicted)}$, $Q_{150(predicted)}/3$). The factor of safety F was defined as the ratio of the measured Q_{150} over Q_d. Since 31 participants predicted the behavior of 5 footings, there was a total 155 values of the factor of safety F. Only once out of 155 was F less than 1, the next worse case was 1.6; the average was 5.4. Therefore it appears that our profession knows how to design spread footings very safely.
7. The settlement S_d under the design load Q_d was read on the measured curves at the value of the predicted design load for each footing and for each participant. The overall average was 10.3 mm which is much smaller than 25 mm. Considering the high factors of safety and the low settlement values, the design load could have been significantly higher. Therefore it appears that our profession could design spread footings more economically.

ACKNOWLEDGEMENTS:

The authors wish to thank all the respondents to this prediction request. They made it possible for our profession to evaluate our current practice and should pride themselves for this important professional contribution. Mr. Al DiMillio, Mr. Mike Adams and the Federal Highway Administration should be thanked for funding and actively participating one more time in an event of great professional interest and benefit. All those who participated in the testing program are most appreciated and acknowledged. Their names and contributions are listed in the preface. Thank you all; it was a lot of fun.

PREDICTING THE SETTLEMENT OF THE TEXAS A&M SPREAD FOOTINGS ON SAND

G. Wiseman (1) and J.G. Zeitlen, (2) Fellow, ASCE

ABSTRACT: A prediction is presented of the load associated with 25 mm and 150 mm settlement for the five square footings on sand that are to be tested prior to the "Settlement 94" conference. Included also is a brief description of the methodology used.

INTRODUCTION

A prediction symposium for spread footings on sand will take place at the time of the forthcoming ASCE Geotechnical Engineering Division specialty conference "Settlement 94" to be held in June 1994 at Texas A&M University. The stated goal of the prediction symposium is to "evaluate the profession's ability to predict behaviour, at small and large deflections, of footings on sand". Ignoring the advice of the sages as recorded in the Talmud about 1800 years ago that "ever since the destruction of the second temple only fools and children engage in prophesying" we indicated our willingness to participate in the prediction symposium. This paper predicts the load associated with 25 mm and 150 mm settlement for five square footings of various sizes.

TEST SITE AND LOADING

The soil at the site is reported to consist of a fine sand extending from the present ground surface to a depth of 11 meters, underlain by a very hard dark gray clay. A water table was observed at a depth of 4.9 meters. About 1.0 meter of overburden had already been removed from the site prior to the site investigation. Five square spread footings founded at a depth of 0.76 m below the present ground surface are to be load tested. They are: 3 m; 3 m; 2.5 m; 1.5 m; and 1 m wide, located within an area of 10 by 15 meters. They are so placed that there may be assumed to be no overlapping of inluence on any particular footing from the loading of adjacent footings.

The footings are to be loaded incrementally at 30 minute intervals. At 30% of the estimated capacity the load will be reduced to zero. At 40% of the estimated capacity the load will be maintained for 24 hours and then loading will continue in increments to 100% of estimated capacity. It is therefore not clear as to whether the 24 hour maintained load will be before or after 25 mm of settlement is first achieved.

SITE INVESTIGATION

The site investigation included a wide range of insitu and laboratory tests. Our methodology for predicting footing behaviour in sands is primarily based on the blow counts in the SPT test. Since the largest footing is 3 m wide we

---

(1) Prof., Dept. of Civ. Engrg., Israel Inst. of Technology, Soil Engineering Building, Haifa 32000, Israel

(2) Prof. Emeritus, Dept. of Civ. Engrg., Israel Inst. of Technology

are mostly concerned with the soil profile to a depth of about 5 meters, i.e. above the water table. From the descriptions provided it would appear that the SPT testing was done at a time and in such a manner that the moisture contents above the water table associated with the SPT testing might well be higher than those to be expected at the time of the load testing.

The micro variability of the site is easily observed from inspecting the 5 Static Cone Penetrometer results. The tip resistances within the depth of interest essentially ranged from about 5 to 10 MPa. However, CPT No. 7, showed locally zero tip resistance at a depth of from 2.2 to 3.0 meters, but an SPT boring at about 1.5 meters distance showed no indication of the existence of such a weak soil layer. There were 6 SPT borings executed at the site. The blow counts ranged from a low of 11 blows/30 cm near the ground surface to 28 blows/30 cm at a depth of 4.9 meters. No clear pattern of spacial variability could be observed. The average SPT results above the water table are summarized below:

TABLE 1 AVERAGE SPT BLOW COUNTS/30 cm

Depth	(m)	0.2	1.0	1.5	2.1	2.8	3.6	4.9
Blows/30 cm	-	11.8	18.8	21.5	16.8	17.2	19.0	22.5

The sands are reported to have less than 20% silt sizes, have a mean size of about 0.2 mm and a uniformity coefficient of about 2. The mineral composition and the angularity have not been described. Based on the SPT and CPT results the conveners of the symposium estimated the Relative Density at 55%. Triaxial specimens prepared at this relative density were reported to show an angle of internal friction (phi) of about 35 degrees.

EVALUATED SOIL PARAMETERS

The sands at the site would appear to be quite similar to those with which we have experience in Israel (Zolkov and Wiseman, 1965). For over 30 years we have been using the USBR overburden corrections for interpreting SPT results in terms of relative density as the first step in predicting the behaviour of foundations in sand profiles (Alpan 1964). This procedure has also now been included in the U.S. Navy Soil Mechanics Design Manual DM-7 in the form of lines of equal relative density (Dr) on a plot which has Y coordinates of vertical effective stress and X coordinates of Blows/30 cm. (See Fig. 3 pg 7.1-87, NAVFAC, 1982). Using this figure, all the blow counts above the water table correspond to relative densities between 70% to over 90%. We assumed the total unit weight of the sand to be 16.0 kN/cu m.

For estimating the angle of internal friction (phi) of the sand, we used the following empirical equation which fits our own data, as well as the test data for this site where a (phi) was reported of 35 degrees for a Relative Density Dr of 55%:

$$\text{phi} = 30 + 9 * [Dr/100] \quad \text{degrees.}$$

SETTLEMENT PREDICTION

Tan and Duncan (1991) evaluated 12 methods of settlement prediction by comparing calculated with measured settlements. Their paper discusses the important concepts of "accuracy" and "reliability". In the design situation we are primarily interested in "reliability" whereas in the context of this prediction symposium our main concern is with "accuracy". The method presented by Alpan (1964) got highest marks for accuracy. The method uses the USBR overburden corrections for interpreting the SPT blow counts in terms of relative density, the coefficient of subgrade reaction from published results for loading tests on 30 cm diameter plates, and the so called Koegler relationship (with a maximum multiplier of 4) for extrapolating to larger footing sizes for computing settlements in the linear range.

The method we have been using for many years for evaluating bearing capacity and settlements at working stresses retains many of the above features and is described below:

1. Obtain the average Relative Density over a depth of 1.5 times the footing width from the SPT values interpreted using charts relating SPT, Relative Density and overburden pressure.
2. Decide on (phi) based on the above Relative Density, then compute the Terzaghi bearing capacity factors and the bearing capacity q(f).
3. Determine the coefficient of vertical subgrade reaction kv(30) for the above Relative Density, using Fig. 6 pg 7.1-219 from NAVFAC (1982)
4. Obtain kv(B) for the actual footing width using:

$$kv(B) = kv(30) * [(B + .3\ m)/(2* B)]^2$$

We have also computed Kv(B) following the procedures proposed by Burland and Burbidge (1985) and they are shown in Table 2, identified as Kv(Burland).

Since we have to predict the load on the various size footings which will cause a settlement of 25 mm and 150 mm it is necessary to assume a functional relationship between load and settlement. We have proceeded as follows:

5. Assume a two parameter hyperbolic relationship for the settlement (s) versus stress (q) relationship. On a plot with (s/q) on the Y-axis and (s) on the X-axis this relationship is a straight line with a slope (m) equal to the inverse of the contact stress (q) at infinite settlement (s) and an intercept (c) equal to the inverse of the coefficient of subgrade reaction at the beginning of loading:

$$(s/q) = c + m * (s)$$

6. Assume that for a loading test on sand there will not be plunging failure and therefore evaluate m in the above equation by assuming

$$1/m = 1.3 * q(f)$$

7. Assume that kv(B) as determined in step (3) applies to a point on the load deformation curve at 40% of the bearing capacity and hence:

$$1/c = kv(B) * (1.3/(1.3-0.4)) = kv(B) * 1.44$$

Using the above procedures the prediction parameters listed in Table 2 were derived:

TABLE 2 PREDICTION PARAMETERS

Width	(m)	1.0	1.5	2.5	3.0
Relative Density	(%)	87	85	82	80
Friction angle	deg	38	37.5	37.4	37
Bearing Capacity	kPa	997	1110	1465	1538
kv(30)	MN/cu m	95	90	85	80
kv(B)	MN/cu m	40	32	27	24
kv(Burland)	MN/cu m	39	27	18	16
1/m	kPa	1296	1443	1905	1999
1/c	MN/cu m	58	47	39	35

The prediction parameters listed above were assumed to be valid for the 30 minute load settlement curve at the site. After examining the results of applying either kv(B) or kv(Burland) as subgrade reaction coefficients, we decided to use Burland's values for generating our predicted stress settlement curves, as furnishing more reasonable results. Predictions are presented below in Table 3.

TABLE 3 TABULATED PREDICTIONS

Footing Width	(m)	No. 1 3 m	No. 2 1.5 m	No. 3 3 m	No. 4 2.5 m	No. 5 1 m
Load @ 25mm (30 min) Q25	(kN)	3900	1300	3900	3000	670
Load @ 150 mm (30 min) Q150	(kN)	11300	2600	11300	7900	1120
Creep @ Q25 (1-30 min)	(mm)	13	-	13	-	-
Total Displ.@ Q25 Year 2014	(mm)	36	-	36	-	-

CONCLUDING REMARKS

Various arbitrary assumptions have been involved in the procedure described above. The end result must of course be examined for reasonableness. This can only be done using well documented case histories for similar soil profiles. No such experience is available to us at this time for the Texas A&M site. It should also be pointed out that we would consider it fortuitous if any loading test results were closer than 30% to the above predictions.

REFERENCES

Alpan, I. (1964). "Estimating the Settlements of Foundations on Sands", Civil Engineering and Public Works Review, November, London, U.K.

Burland, J.B., and Burbidge, M.C. (1985). "Settlement of Foundations on Sand and Gravel", Proc. Inst. of Civil Engrs., (Part 1), London, 1325-1381.

NAVFAC (1982), Soil Mechanics - Design Manual 7.1, U.S. Dept. of Navy, U.S. Govt. Printing Office, Washington, D.C. 348 p.

Tan, C.K., and Duncan, J.M. (1991). "Settlement of Footings on Sand, Accuracy and Reliability", ASCE Geotechnical Congress, Geotechnical Special Publication No. 27. Vol. 2, 446-455.

Zolkov, E., and Wiseman, G. (1965). "Engineering Properties of Dune and Beach Sands and the Influence of Stress History." Proc. 6th Int. Conf. on Soil Mechanics and Foundation Engineering, Montreal, Canada, Vol. 1, 134-138.

CLASS A PREDICTION OF SHALLOW FOOTING SETTLEMENTS

by

HARRY G. POULOS F.ASCE
Senior Principal, Coffey Partners International Pty Ltd
12 Waterloo Road, North Ryde, Australia 2113; also
Professor Civil Engineering, University of Sydney

INTRODUCTION

This paper presents "Class A" predictions of the settlement behaviour of five shallow footings on sand, and has been prepared for the ASCE Specialty Conference, Settlement '94. It outlines the approach adopted and presents comparisons with alternative approaches. The requested predictions are summarized in tabular form.

THE PREDICTION EXERCISE AND AVAILABLE DATA

Five square footings are to be tested at a site on the Riverside campus of Texas A&M University. The subsoil profile consists essentially of sandy soils to a depth of about 10m, underlain by very hard clay. Figure 1 shows a simplified representation of the subsoil profile. The report made available to predictors includes the results of 6 borings, and a variety of in-situ data including SPT, CPT, PMT, DMT, step-blade, borehole shear and cross-hole seismic tests. Laboratory predictors have been requested to make the following predictions:

1. load for 25mm of settlement after 30 minutes,
2. load for 150mm of settlement,
3. creep settlement between 1 and 30 minutes,
4. settlement in the year 2014 under the load corresponding to 25mm settlement.

The first two predictions are required for all five footings and the last two for only the two largest footings.

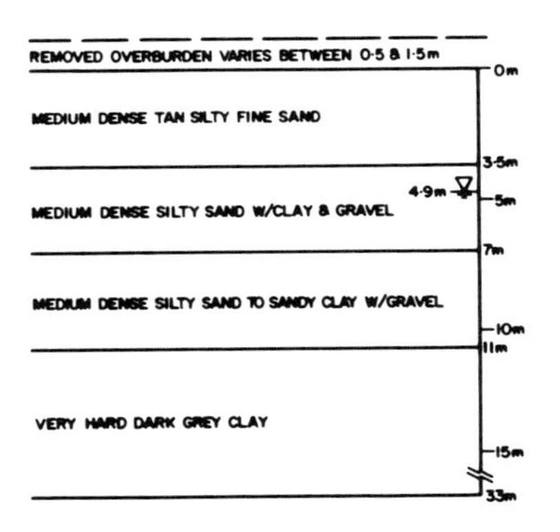

Figure 1 Simplified Subsoil Profile

PREDICTION APPROACH ADOPTED

A reasonably practical approach has been adopted, rather than an analytical approach based on proper constitutive modelling of the sand behaviour. It follows broadly the procedure described by Oweis (1979) and involves the following steps:

1. calculation of the small-strain Young's modulus of the soil at various depths below the footing by correlation with the SPT data.
2. calculation and summation of vertical strains below the footing (assumed rigid) via elastic theory.
3. allowance for the reduction of Young's modulus as the strains increase, using the resonant column data as a guide.

As incremental analysis has been used, with increments of about 10 kPa in applied footing pressure being used. A computer program, SAFS (Simplified Analysis of Footing Settlement) has been written to evaluate this analysis.

In addition to the SAFS analysis, several alternative methods have been considered for one of the footings, in an attempt to avoid gross mis-prediction. These include the methods of Terzaghi and Peck (1967), Burland and Burbridge (1985), Schultze and Sherif (1973), and linear elastic layer methods using values of secant Young's modulus E_s related either to SPT (E_s = 2N MPa) or to CPT (E_s = 3.5 q_c, where q_c = cone resistance). In all these calculations, Poisson's ratio of the soils has been taken as 0.3, while (where appropriate) the coefficient of earth pressure at rest K_o has been assumed to decrease from 1.0 beneath the footings to 0.4 at a depth of about 6m. Attempts were made to derive K_o from DMT tests, but consistent results were not obtained.

Estimates of the ultimate bearing capacity have been made using the solutions presented by Vesic (1975), using the laboratory triaxial results to assess the friction angle of the soils. The computed load-settlement curves from the SAFS analysis, have been "cut-off" at the estimated bearing capacity.

Finally, time effects in the longer-term have been estimated extremely crudely, in the absence of appropriate data in the prediction package, using the approach suggested by Schmertmann (1970). Judgement has been employed to assess that the time-dependent settlement between 1 and 30 minutes will be less than 1mm at the load level corresponding to 25mm settlement at 30 minutes.

COMPARATIVE PREDICTIONS

For one of the 3m x 3m footings (Footing 3), Table 1 shows various predictions of the load Q_{25} for 25mm settlement. The range of Q_{25} values is 3210 kN to 5890 kN, with the prediction from the SAFS analysis lying within this range. The largest predictions are given by the Schultze and Sherif method, while the smallest are from the elastic method using E_s = 3.5 q_c, and the Terzaghi and Peck approach, which is not a prediction method, but a design method.

Figure 2 compares the predicted load-settlement curve from SAFS for a 1m square footing with that presented by Burland and Burbridge (1985) for a sand with an SPT value between 16 and 25. The agreement is quite reasonable and suggests that the general shape of the load-settlement curve given by SAFS is reasonable.

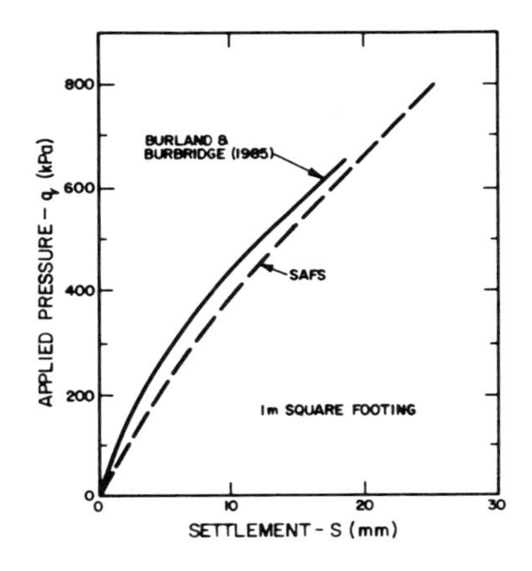

Figure 2 Comparison Between Load-Settlement Curves

SUMMARY OF PREDICTIONS

Table 2 summarizes the predictions requested. The prediction for Footing 4 is for the as-built dimensions. As mentioned above, the predictions of time-dependent movement are based largely on subjective assessment, since there is no directly relevant data in the prediction package provided.

TABLE 1
COMPARISON BETWEEN VARIOUS PREDICTIONS FOR Q_{25} LOAD FOR FOOTING 3

Method	Average Pressure for 25mm Settlement (kPa)	Q_{25} kN
Terzaghi and Peck (1967)	369	3320
Schultze and Sherif (1973)	654	5890
Burland and Burbridge (1985)	619	5570
Linear Elastic (using E_s = 2N MPa)	552	4970
Linear Elastic (using E_s = $3.5q_c$)	357	3210
SAFS	410	3690

TABLE 2
SUMMARY OF PREDICTIONS

	Footing 1 3m x 3m	Footing 2 1.5m x 1.5m	Footing 3 3m x 3m	Footing 4 2.5m x 2.5m	Footing 5 1m x 1m
Load for 25mm of settlement Q_{25} on the 30 minute load settlement curve (kN)	2790	5040	3690	2560	800
Load for 150mm of settlement Q_{150} on the 30 minute load settlement curve (kN)	12690	11340	14580	9680	1120
Creep settlement between 1 minute and 30 minutes for Q_{25}, Δs(mm)	< 1	-	< 1	-	-
Settlement in the year 2014 under Q_{25} (mm)	37	-	37	-	-

CONCLUSIONS

A very comprehensive set of geotechincal data has been obtained for the test site, using both in-situ and laboratory tests. The author has used the SPT data as the primary source of data for his predictions, although use has also been made of CPT, DMT tests, and laboratory triaxial and resonant column tests. The settlements have been calculated via an approach similar to that developed by Oweis (1979). Despite this welter of data, there appears to be a lack of data related to initial in-situ stresses and time-dependent strains. Consequently, a significant degree of judgement has been applied in order to obtain the necessary predictions.

REFERENCES

Burland, J.B. and Burbridge, M.C. (1985). "Settlement of Foundations on Sand and Gravel." Proc. Instn. Civ. Eng. Pt. 1, 78, 1325-1381.

Oweis, I.S. (1979). "Equivalent Linear Model for Predicting Settlement of Sand Bases." J. Geotech. Eng. Divn, ASCE, 105, GT12, 1525-1544.

Schultze, E. and Sherif, G. (173). "Prediction of Settlements from Evaluated Settlement Observation for Sand." Proc. 8th Int. Conf. Soil Mechs. Foundn. Eng. Vol. 1.3 225-230.

Terzaghi, K. and Peck, R.B. (1967). "Soil Mechanics in Engineering Practice." John Wiley, New York.

Veni, A.S. (1975). "Bearing Capacity of Shallow Foundations." Ch. 3 of Foundation Engineering Handbook, Ed. H.F. Winterkorn and H. Fang, Van Nostrand Reinhold, New York.

Settlement prediction of spread footing based on laboratory test results

Mohammed S. A. Siddiquee[1], Tadatsugu Tanaka[2] and Fumio Tatsuoka[3]

ABSTRACT: A prediction of spread footings of different sizes, based on a very simple elasto-plastic isotropic hardening (frictional) model is described. The parameters used in the model was derived from the conventional TC tests results along with resonant column test results as the small strain complementary counterpart, provided by the moderator. The depth in the ground modelled in the analysis was determined from the CPT and borelogs observation. The model is a pressure sensitive one.

INTRODUCTION

Laboratory based predictions usually need relatively large number of tests to attain a certain amount of reliability. In this case, total six TC tests and four Resonant Column tests were performed. As the number of tests are not significant (especially TC tests) and they were performed on reconstituted samples, this introduces a unavoidable and undesirable uncertainty in the prediction. Anyway, the model used here is a rate independent material model, so time effects are excluded in the analysis. Effect of localization is not considered here as the stress level did not reach the peak stress.

MODEL DESCRIPTION

This is a elasto-plastic isotropic hardening model with non-associated flow behaviour. Elastic part is modeled using the formulation by Iwasaki and Tatsuoka (1977). The plastic part is modelled by non-associated flow rule. So a generalized Mohr-Coulomb yield function is used along with a Drucker-Prager type plastic potential. The original stress-dilatancy relation (Rowe, 1962) was modified by introducing a mobilized dilatancy equation . The yield function is given by

$$\Phi = \alpha I_1 + \frac{1}{g(\theta)}\sqrt{J_2} - K = 0 \qquad \textbf{(1)}$$

where,

$$\alpha = \frac{2\sin\varphi_{mob}}{\sqrt{3}\left(3 - \sin\varphi_{mob}\right)}, g(\theta) = \frac{3 - \sin\varphi_{mob}}{2\sqrt{3}\cos\theta - 2\cos\theta\sin\varphi_{mob}}, \qquad \textbf{(2)}$$

I_1 = 1st stress invariant, J_2 = 2nd invariant of deviatoric stress,

K = Cohesion term (not used here), θ = Lode angle

The plastic potential is defined as follows

$$\Psi = \alpha' I_1 + \sqrt{J_2} - K = 0 \qquad \textbf{(3)}$$

[1]Graduate student, IIS, The University of Tokyo, Japan
[2]Professor, Meiji University, Japan
[3]Professor, IIS, The University of Tokyo, Japan

where,

$$\alpha' = \frac{2 \sin \psi}{\sqrt{3}(3 - \sin \psi)}, \quad \sin \psi = \frac{\sin \varphi_{mob} - \sin \varphi_r}{1 - \sin \varphi_{mob} \sin \varphi_r} \tag{4}$$

φ_{mob} = Mobilized angle of internal friction ,

φ_r = Residual angle of internal friction

The growth of the yield function is modeled by a power function with a flat peak (Tanaka and Kawamoto, 1988) as shown below.

$$\alpha(\kappa) = R_{max}\left(2\frac{\sqrt{\kappa \varepsilon_f}}{\kappa + \varepsilon_f}\right)^m \tag{5}$$

where,

$$k = \int d\bar{\varepsilon}_p, \quad R = \frac{\sigma_1}{\sigma_3} \tag{6}$$

$$d\bar{\varepsilon}_p = \left[2\left(de_{11}\right)^2 + 2\left(de_{22}\right)^2 + 2\left(de_{33}\right)^2 + \left(d\gamma_{12}\right)^2 + \left(d\gamma_{23}\right)^2 + \left(d\gamma_{31}\right)^2\right]^{\frac{1}{2}} \tag{7}$$

$de_{ij} \sim d\gamma_{ij}$ are the deviatoric components of plastic strain, ε_f= plasticity parameter, κ at the peak stress state.

PARAMETER DETERMINATION

The model described above needs parameters to fit to the materials under the spread footings. For the elastic parameters, shear modulus is modeled using the following equation (2)

$$G = G_0 \frac{(2.17 - e)^2}{1 + e}(\sigma_m)^n \tag{8}$$

Using the tabulated tests data presented in Table 8 of the material report (Gibbens and Briaud, 1994), above Eq. (9) was fitted on a log-log plot to get the following tabulated results

Soil type	Depth (m)	Constant G_0 (MPa)	Exponent n
Silty sand	0	11.96	0.328
Clean sand	1.6	8.12	0.594
Clean sand	3.3	8.24	0.657
Clayey sand	6.0	5.47	0.324

Table 1 Determination of the elastic parameters.

From Table 1, it is clear that the there is no clear tendency of change with depth, so the values are averaged to get the parameters as G_0=8.45 MPa and n=0.476. Poisson's ratio is assumed as 0.3 due to the lack of small strain data.

For the plastic part of the formulation, it is necessary to fit the growth function of the isotropic hardening model. The growth function (Eq. 5) is based on the norm of plastic component of the deviatoric strains. When κ in Eq. (6) is evaluated with triaxial strain variables, it reduces to maximum shear strain; $\kappa=(\varepsilon_a-\varepsilon_r)$. The TC stress-strain, volumetric strain data supplied in Fig. 12 through 15 of the material report (1) are used to get the maximum shear and then a nonlinear fitting is used to fit to Eq. 5.

	0.6 m depth samples			3.0m depth samples		
Conf. press.(kPa)	34.0	138.0	345.0	34.0	138.0	345.0
γ_f	0.124	0.265	0.213	0.100	0.243	0.343
Exponent m	0.639	0.794	0.602	0.838	0.707	0.630

	0.6 m depth samples	3.0m depth samples
γ_f	$=0.193 + 0.066 \log(\sigma_c)$	$=0.214 + 0.160 \log(\sigma_c)$
Exponent m	$=0.679 - 0.012 \log(\sigma_c)$	$=0.737 - 0.207 \log(\sigma_c)$

Table 2 showing the parameters for the hardening function.

The maximum stress ratios (R_{max}) are calculated from Table 7 of the material report (1). It is seen that in case of 0.6m samples, data have no definite tendency, and when regressed linearly, shows a increasing tendency with confining pressure which is very unreliable. Similar modeling with Toyoura sand (Tatsuoka and Siddiquee *et. al.*, 1993) always showed decreasing tendency with confining pressure. Samples from 3.0m depth showed a decreasing curve. In the FEM analysis strength varied according to the depth. The equations are

$$R_{max} = 3.883 \qquad 0.0\text{m} \le d \le 1.2\text{m}$$

$$R_{max} = 4.106 - 0.275 \cdot \log_{10}(\sigma_c) \quad 1.2\text{m} < d \le 10\text{m} \qquad \textbf{(10)}$$

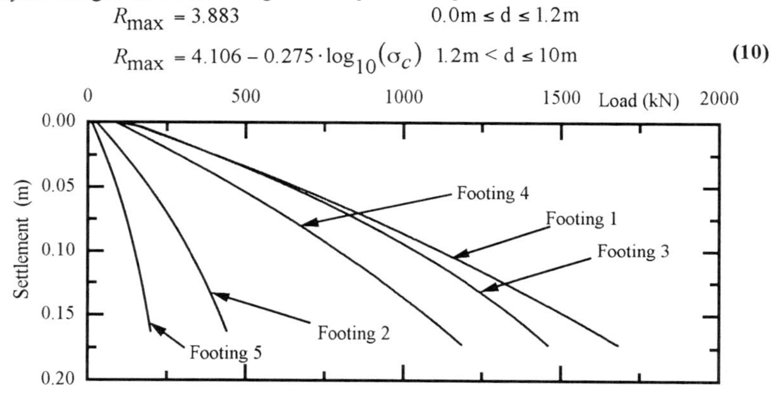

Fig.1 shows the load-displacement curves

RESULTS AND DISCUSSION

Footings were discretized by identical mesh of 1753 nodes and 1347 elements of 8-noded brick elements with 1-point integration. Different mesh sizes were simulated by simple affine scaling in x and y direction. The depth of the analysis was kept fixed around 10 m from the SPT tests data. The non-linear solutions were obtained by a highly optimized Dynamic Relaxation (DR) solver (5). The load-settlement curves of footings are shown in Fig. 1. Results of footing 3 and 1 are different, because footing 3 is predicted by introducing a weak layer at a depth of 2m to 3.5m. The thickness and a reduction factor is determined from Fig. 27 (1) and Fig.37 (1) respectively. A reduction factor of 0.675 is used to reduce the R_{max} (Eq. 10) in that layer.

	Footing 1 3m x 3m	Footing 2 1.5 x 1.5	Footing 3 3m x 3m	Footing 4 2.5m x 2.5m	Footing 5 1m x 1m
Load for 25 mm of settlement Q25 on the 30 minute load settlement curve (kN)	415.38	116.22	406.5	294.6	58.52
Load for 150 mm of settlement Q150 on the 30 minute load settlement curve (kN)	1502.46	421.57	1325.7	1086.33	199.83
Creep settlement between 1 minute and 30 minute for Q25, Δs (mm)	NA		NA		
Creep settlement between 1 minute and 30 minute for Q25, Δs (mm)	NA		NA		

CONCLUSIONS

The predictions are based on the available data. There are variations in the data which can reflect in the variation of the prediction. Moreover, the interaction of the footings are not considered in the prediction.

REFERENCES

1. Gibbens R. and J.-L. Briaud, "Data and prediction request for the spread footing prediction event sponsored by FHWA at the occasion of the ASCE specialty conference : Settlement '94", a report published from Texas A & M University, 1993.

2. Iwasaki, T. and Tatsuoka, F., "Effects of grainsize and grading on dynamic shear moduli of sands", Soils and Foundations, vol. 17, No. 3, pp. 19-35.

3. Rowe, P. W., "The Stress Dilatancy Relation for Static Equilibrium of an Assembly of Particles in Contact", in Proceeding of Royal Society, London, 1962, series A, pp. 500-527.

4. Tatsuoka F., Siddiquee ,Mohammed S. A., Park C.-S., Sakamoto M. and Abe F., "Modeling stress-strain relations of sand", Soil and Foundations, Vol. 33, No. 2, pp. 60-81, June 1993.

5. Tanaka T. and Kawamoto O.,"Three dimensional finite element collapse analysis for foundations and slopes using dynamic relaxation" in Proceedings of Numerical methods in geomechanics, Insbruch, pp. 1213-1218 (1988)

LOAD - SETTLEMENT PREDICTIONS FOR SPREAD FOOTING EVENT

By Vincent Silvestri[1], Member, ASCE

ABSTRACT: Based on the hyperbolic stress-strain model and Vesic's bearing capacity theory, load-settlement predictions are given for each shallow footing. Initial values of Young's modulus were estimated from CPT, DMT and PMT field results.

INTRODUCTION

The final usefulness of any particular constitutive relationship is dictated in large part by a proper balance among a) the versatility of the theory to characterize experimental data obtained from a variety of different tests, b) the ability of the resulting relationship to predict behavior for conditions other than those which were used to calibrate the model, and c) the ease with which the formulation can be adapted to the solution of practical problems. Exercises in fitting data and predicting response patterns therefore provide valuable comparisons among different theories and serve to identify and clarify the relative merits of each.

Because of the difficulty and expenses associated with obtaining undisturbed samples of granular soils, few settlement estimates are based on the results of laboratory tests. Empirical methods based on correlations between observed settlements and the results of field tests are generally used to estimate the settlement of footings founded on granular deposits. Commonly used field tests are the standard penetration test (SPT), the static cone penetration test (CPT), the plate load test, the flat dilatometer test (DMT), and the pressuremeter test (PMT). Methods involving elastic theory have been suggested as a means of predicting settlements of footings on cohesionless soils (Bowles 1988). Such methods apparently have had limited applications, the limitation being caused by the difficulty of evaluating the in-situ stress-strain properties.

In the present case, a variety of both in-situ and laboratory tests were carried out in order to determine the constitutive properties of the granular soil deposit. On the basis of the results obtained from these tests, each layer of the soil deposit was characterized by a nonlinear stress-strain curve and load-settlement relationships were determined for each footing using the theory of elasticity.

PREDICTIONS

Because the cohesionless soils encountered at the site behave as nonlinear materials, their stress-strain relationships were represented by means of the hyperbolic model (Kondner 1963). As the first parameter employed in this model is the initial tangent modulus E, its value was determined at each depth by means of the results obtained from a) the cone penetration

---

[1] Professor of Civil Engineering, Ecole Polytechnique, P.O. Box 6079, Station "Centre-ville", Montreal, Qc., Canada H3C 3A7

tests, using the relation $E = 3.5\ q_c$ (Bowles 1988; Robertson 1991), where q_c = point resistance, b) flat dilatometer tests, using the relation $E = (1 - \nu^2)\ E_d$ (Marchetti 1980; Schmertmann 1988), where ν = Poisson's ratio and E_d = dilatometer modulus, and c) the pressuremeter tests, using either the relation $E = 3\ E_o$ or $E = E_r$ (Baguelin et al. 1978), where E_o = pressuremeter modulus and E_r = reload modulus.

For the determination of the vertical strain distribution in the soil deposit, the theory of elasticity was used. At any depth z below the centerline of the footing, the vertical strain ϵ_z is given by

$$\epsilon_z = \frac{\Delta\sigma_z}{E} - 2\nu\ \frac{\Delta\sigma_x}{E} \tag{1}$$

where $\Delta\sigma_z$ = vertical stress increase; $\Delta\sigma_x$ = horizontal stress increase; E = Young's modulus and ν = Poisson's ratio.

The stress increases $\Delta\sigma_z$ and $\Delta\sigma_x$ where calculated using the following equations (Poulos and Davis 1974):

$$\Delta\sigma_z = \Delta p \cdot I_0 \tag{2}$$

and

$$\Delta\sigma_x = \Delta p \left[I_2 - (1 - 2\nu)\, I_2'\right] \tag{3}$$

where Δp = pressure applied to footing; I_o, I_2, I_2' = influence factors. Poisson's ratio ν was determined by using the axial strain-volumetric strain relationships obtained in the two series of triaxial tests, performed at confining pressures of 34, 138 and 345 kPa, respectively. The expected settlement was found by summing up the vertical strains calculated in the soil deposit along the centerline axis of each footing.

As the second parameter employed in the hyperbolic model is the maximum compressive stress, its value was determined using Vesic's bearing capacity theory (Vesic 1975) and the results are shown in Table 1. Please note that the friction angle ϕ of the granular soil was assumed to be equal to 36°, on the basis of either the relationship between q_c and ϕ (Meyerhof 1974) or that between the standard penetration blow count N and ϕ (Terzaghi and Peck 1967). However, in the latter case, measured N values were adjusted for overburden pressure, using the correction factor

$$C_N = 0.77 \log \frac{2000}{\sigma'_{vo}} \tag{4}$$

where σ'_{vo} = vertical effective stress at test level.

Table 1. Bearing capacities of footings

Footing width, B (m)	1	1.5	2.5	3
Bearing capacity (kPa)	1104	1229	1392	1484
Failure load (kN)	1104	2765	8700	13356

For the determination of the load-settlement relationships, the values of the initial Young's moduli mentioned above were considered to represent 1 minute values. In order to determine the load at a settlement of either 25 mm or 150 mm on the 30 minute load-settlement curve, appropriate Young's moduli were calculated using the results of the pressuremeter tests in which the pressures were held constant for a duration of 10 minutes. These tests allowed to calculate also the creep settlement between the 1 minute and the 30 minute reading, for each footing for the Q25 load, and as well as the total settlement for the 3 m X 3 m footing in the year 2014.

Table 2. Load-Settlement Predictions

Item	Footing 1 3 m x 3 m	Footing 2 1.5 m x 1.5 m	Footing 3 3 m x 3 m	Footing 4 2.5 m x 2.5 m	Footing 5 1 m x 1 m
Load for 25 mm of settlement Q25 on the 30 minute load settlement curve (kN)	1929	771	1929	1488	448
Load for 150 mm of settlement Q150 on the 30 minute load settlement curve (kN)	6857	2143	6857	5060	1085
Creep settlement between 1 minute and 30 minute for Q25, Δs (mm)	1.3		1.3		
Settlement in the year 2014 under Q25 (mm)	31		31		

All load-settlement predictions are summarized in Table 2. A comparison between the loads shown in this table for the 150 mm settlement indicates that there exists the possibility that the 1 m x 1 m footing reaches failure.

ACKNOWLEDGEMENTS

The author expresses his deep gratitude to the National Research Council of Canada and the Department of Education of Quebec (FCAR) for the financial support received in the course of this study. The author wishes to thank the sponsors and organizers of this symposium for the opportunity to present his solution and to participate as predictor. The problems posed for the predictors have stimulated the thinking of the author about the heterogeneity of the soil deposit and the variety of test results obtained from the soil investigation. The author expects that many valuable lessons will be learned as a result of the exercises performed for the symposium.

REFERENCES

Baguelin, F., Jézéquel, J.F. and Shields, D.H. (1978). The Pressuremeter and Foundation Engineering. Trans Tech Publications, Clausthal, Germany.

Bowles, J.E. (1988). Foundation Analysis and Design. Fourth Edition, McGraw-Hill Book Company, New York, N.Y.

Kondner, R.L. (1963). Hyperbolic Stress-Strain Response: Cohesive Soils. J. Soil Mech. Found. Div., ASCE, 80(1), pp. 115-143.

Marchetti, S. (1980). In Situ Tests by Flat Dilatometer. J. Geot. Engrg. Div., ASCE, 103(3), pp. 299-321.

Meyerhof, G.G. (1974). General Report: Outside Europe. Proc. ESOPT, Stockholm, Vol. 2.1, pp. 40-48.

Poulos, H.G. and Davis, E.H. (1974). Elastic Solutions for Soil and Rock Mechanics. John Wiley & Sons Inc., New York, N.Y.

Robertson, P.K. (1991). Estimation of Foundation Settlements in Sand from CPT. Proc. Geotech. Engrg. Cong. 1991. ASCE Geotechnical Special Publication No. 27, Vol. II, pp. 764-775.

Schmertmann, J.H. (1988). Guidelines for Using the CPT, CPTU and Marchetti DMT for Geotechnical Design. Vol. III - DMT Test Methods and Data Reduction. U.S. Dept. of Transportation, Report No. FHWA - PA - 024 + 84-24.

Terzaghi, K. and Peck, R.B. (1967). Soil Mechanics in Engineering Practice. John Wiley & Sons Inc., New York, N.Y.

Vesic, A.S. (1975). Bearing Capacity of Shallow Foundations. In "Foundation Engineering Handbook", H.F. Winterkorn and H.Y. Fang, Editors. Van Nostrand Reinhold Company, New York, N.Y.

Estimation of Spread-Footing Settlement on a Sand Subgrade

John S. Horvath[1]

INTRODUCTION

This paper presents before-the-fact settlement estimates made by the writer for five spread footings constructed at a National Geotechnical Experimentation Site located at the Texas A&M University Riverside Campus. These footings will be test loaded as part of the *Settlement '94* ASCE specialty conference to be held in June 1994, with the results to be published subsequently. Also presented in this paper are brief descriptions of the analytical methods used to make these estimates and the site-characterization procedures for the development of parameters used in the analyses.

The only data used in developing analytical parameters were those provided to all participants (Gibbens and Briaud 1993) plus various supplements and clarifications to this reference that were provided subsequently.

ANALYTICAL METHODS

Load for 25 mm Settlement (Q_{25})

A settlement of this magnitude is between 0.8% and 2.5% of the footing widths analyzed. The basic procedure used was the strain-distribution method discussed in Schmertmann (1978). An important modification was a separation of settlement into a reload component within the overconsolidated (OC) range and a virgin-curve component within the normally consolidated (NC) range. Conceptually, this corresponds to the procedure proposed by Leonards and Frost (1988) with the important exception that the CPTU data only (not combined DMT and CPT as used by Leonards and Frost) were used by the writer.

Load for 150 mm Settlement (Q_{150})

A settlement of this magnitude is between 5% and 15% of the footing widths analyzed. This was judged to be within the range of a bearing-capacity failure (Vesic 1975). Therefore, estimates of Q_{150} were based on bearing-capacity analyses. The difficulty was in choosing an appropriate friction angle to use in the classical bearing capacity equations because of the sensitivity of the results to this parameter (Vesic 1975). Use of peak friction angles was judged to result in unreasonably high calculated values for the bearing capacity. Consequently, the constant-volume friction angle was used and the resulting bearing capacities increased arbitrarily by 50% to account for peak strengths.

Creep

Time-dependent settlements under the Q_{25} load, referred to as Δs_1 and

---

[1]Associate Professor, Manhattan College, Civil Engineering Department, Bronx, New York 10471, U.S.A.

s_{2014}, were estimated based on a model of a reduced Young's modulus with time (as opposed to keeping the modulus constant and using an empirical increase in settlement as recommended in Schmertmann (1978)). The magnitude of the modulus reduction was based on an average of the relaxation data obtained in the PMT tests.

SITE CHARACTERIZATION

In-Situ Stress State

A key element in calculating Q_{25} using the previously described procedure is to calculate the existing overburden stress, $\bar{\sigma}_{v_o}$, and estimate the maximum past effective (yield) stress, $\bar{\sigma}_{v_m}$. This was limited to a depth of approximately 7 m which was the maximum depth of interest used in any of the analyses. Overburden stresses were based on the assumptions that ground water was at a depth of 4.9 m (based on observation well data), the total unit weight above ground water, γ_t = 17.4 kN/m^3, and the buoyant unit weight, γ_b = 9.3 kN/m^3. The total unit weight was based on water contents measured in the SPT samples.

Yield stresses were estimated individually for each footing using the CPTU data and the following equations (p_a = atmospheric pressure)

$$D_r^2 = \frac{(q_c/p_a)}{305\,OCR^{0.18}\,(\bar{\sigma}_{v_o}/p_a)^{0.5}} \tag{1}$$

$$\bar{\sigma}_{h_o} = p_a \frac{(q_c/p_a)^{1.25}}{35\,e^{(D_r/20)}} \tag{2}$$

$$K_o = \bar{\sigma}_{h_o}/\bar{\sigma}_{v_o} \tag{3}$$

$$K_o = (1-\sin\phi_{cv})\,OCR^{\sin\phi_{cv}} \tag{4}$$

$$\bar{\sigma}_{v_m} = OCR\,\bar{\sigma}_{v_o} \tag{5}$$

Eqs.1 to 4, inclusive, were solved iteratively until satisfactory convergence (based on a change in OCR between iterations) was obtained. The final value of OCR was then used in Eq.5. Eqs.1 and 2 are empirical correlations given in Kulhawy and Mayne (1990). Eq.4 is from Mesri and Hayat (1993). Space precludes presentation of the results here. However, overconsolidation effects were judged to be significant (approaching a state of Rankine passive failure near the surface) so that a significant portion of the calculated settlements occurred within the reload (OC) range.

Young's Modulus

The Young's moduli in the NC and OC ranges were estimated in a two-step process by first estimating the drained constrained moduli in the NC and OC ranges using the following empirical correlations for CPT data from Kulhawy and Mayne (1990)

$$M_{nc} = q_c 10^{(1.09-0.0075 D_r)} \tag{6}$$

$$M_{oc} = q_c 10^{(1.78-0.0122 D_r)} \tag{7}$$

The correlation between constrained and Young's moduli was made using

$$E = M \frac{(1+\nu)\,(1-2\nu)}{(1-\nu)} \tag{8}$$

Eq.8 is from linear-elastic theory. A Poisson's ratio of 0.35 was assumed.

Angle of Internal Friction

The peak angle of internal friction, ϕ_{peak}, was estimated using the following empirical correlation for CPT data from Kulhawy and Mayne (1990)

$$\phi = 17.6 + 11.0 \log\left[\frac{(q_c/p_a)}{(\bar{\sigma}_{v_o}/p_a)^{0.5}}\right] \tag{9}$$

Results within a depth B below foundation level for each footing were relatively high, in the range approaching 50°. The constant-volume friction angle, ϕ_{cv}, was assumed to be 35° for all footings based on the lab triaxial results.

RESULTS

Table 1 summarizes the estimated settlements calculate using the procedures described previously.

	Footing 1	Footing 2	Footing 3	Footing 4	Footing 5
Q_{25} (kN)	4500	1450	4500	3125	900
Q_{150} (kN)	20000	4200	23000	13500	1650
Δs_1 (mm)	1		1		
s_{2014} (mm)	10		10		

TABLE 1. Summary of Estimated Settlements

As a general comment (the validity of which can only be evaluated after the footings are tested and the results published), it would appear that the primary research effort for improving analytical methods for the settlement of shallow foundation on coarse-grained soils should focus on developing reliable techniques for estimating the yield stress in such soils so that the distribution of strains between NC and OC regions can be properly considered. As a secondary research task, a rational method for choosing an appropriate peak friction angle for bearing-capacity analysis needs to be developed.

APPENDIX. REFERENCES

Gibbens, R. and Briaud, J.-L. (1993). "Data and Prediction Request for the Spread Footing Prediction Event Sponsored by FHWA at the Occasion of the ASCE Specialty Conference: Settlement '94." Texas A&M University, College Station, Tex., U.S.A., 72 pp.

Kulhawy, F H. and Mayne, P. W. (1990). "Manual on Estimating Soil Properties for Foundation Design." *Final Report - Electric Power Research Institute Project 1493-6*, Cornell University, Geotechnical Engineering Group, Ithaca, N.Y., U.S.A.

Leonards, G. A. and Frost, J. D. (1988)."Settlement of Shallow Foundations on Granular Soils." *Journal of Geotechnical Engineering*, ASCE, 114(7), 791-809.

Mesri, G. and Hayat, T. M. (1993). "The Coefficient of Earth Pressure at Rest." *Canadian Geotechnical Journal*, 30(4), 647-666.

Schmertmann, J. H. (1978). "Guidelines for Cone Penetration Test Performance and Design." *Report No. FHWA-75-78-209*, U.S. Department of Transportation, Federal Highway Administration, Washington, D.C.

Vesic, A. S. (1975). "Bearing Capacity of Shallow Foundations." Chapter 3, *Foundation Engineering Handbook*, H. F. Winterkorn and H.-Y. Fang (eds.), Van Nostrand Reinhold, New York, N.Y., U.S.A.

SPREAD FOOTING PREDICTION EVENT AT THE NATIONAL GEOTECHNICAL EXPERIMENTATION SITE ON THE TEXAS A&M UNIVERSITY RIVERSIDE CAMPUS

DAVID THOMAS [1]

ABSTRACT

Ground conditions at the National Geotechnical Experimentation Site are outlined. Descriptions of methods of analysis which have been employed to estimate the settlements of shallow spread footings constructed at the site are given. Predictions are made of loads to cause predetermined vertical settlements as well as estimates of times required for these movements to occur.

INTRODUCTION

This paper describes the procedures to predict the settlement of spread footings having different dimensions constructed and loaded at the National Geotechnical Experimentation site at the Texas A&M University Riverside Campus.

The estimates of settlement are based exclusively on Cone Penetration Test (CPT) results. Procedures and the assumptions made in the formulation of the analyses are described as are the methods for interpreting the CPT profiles and prediction of times for footing settlements to occur.

As requested of participants, only vertical settlements have been estimated.

SUB-SURFACE CONDITIONS

In the preamble to test data provided, ground conditions have been generally described as :-

0.0 to 3.5 m	Medium dense tan silty fine sand
3.5 to 7.0 m	Medium dense silty sand with clay and gravel
7.0 to 11.0 m	Medium dense silty sand to sandy clay with gravel
11.0 to 33.0 m	Very hard dark grey clay

Ground water stood at 4.9 m below ground level.

Interpretation of CPT profiles (Thomas,1964), with the exception of CPT-7, confirms these conditions. At CPT-7, the indication is that from a depth of approximately 2.25 to 3.00m below ground level the penetrometer encountered a layer of soft clay. The assessed average cone resistance of this layer is 0.32 MPa which would correspond to an undrained shear strength of about 100kPa (Thomas, 1965).

1 - Senior Lecturer, Dept. of Civil Eng. and Building, Central Queensland University, Rockhampton, Queensland, 4702, Australia

PREDICTION RATIONAL

Estimation of footing settlement is based on the assumption that the footing pressure (q,kPa) / footing settlement (ρ,mm). follows a hyperbolic curve having initial and final assymptotes corresponding to an initial tangent modulus (1/A)and an "ultimate bearing capacity" (1/BS) respectively. Hyperbolic stress - strain behaviour for sands was initially proposed by Kondner and Zelasko (1963) and supported by Chin (1983). Further interpretation of this relationship has been considered in detail and related to the settlement of spread footings by Thomas (1993).

The settlement prediction equation is expressed graphically in Figure 1 :-

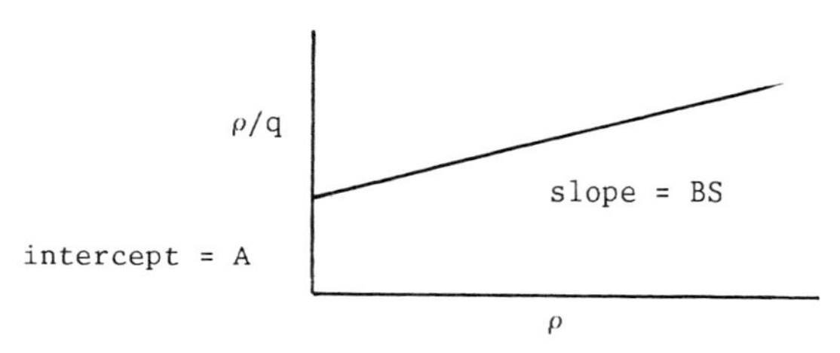

Figure 1

Consequently, settlement and pressure are related by :-

$$\rho = A*q/(1-BS*q) \qquad \text{.................(1)}$$

or

$$q = \rho/(A+BS*\rho) \qquad \text{.................(2)}$$

The initial tangent modulus is obtained from work carried out by Thomas (1966 & 1968) which gave a relationship between cone resistance (qc),Mpa and compression parameter $E/(1-\nu^2)$, kPa. Where E is the tangent modulus and ν is Poisson's ratio. The mathematical relationship (Thomas,1993) has been determined as :-

$$(1-\nu^2)/E = 4.1E{-}06*(70.0/qc)^{0.67} \qquad \text{......(3)}$$

Estimations of settlement for homogeneous, isotropic elastic materials may be calculated from the Schleicher (1926) expression :-

$$1/A = (1-\nu^2)/E*B*I_{se}*I_{de} \qquad \text{...........(4)}$$

where B is the width of footing and I_{se} and I_{de} are influence factors for shape and depth and may be determined from Giroud (1968) and Fox (1948) respectively.

However, it has been demonstrated by several workers Bjerrum & Eggestad (1963),Brown & Gibson (1972) and Parry (1978) that where the elastic modulus increases with depth, equation (4) requires to be modified. Thomas (1993) has demonstrated that to allow for this variation, equation (4) may be changed to :-

$$1/A = (1-\nu^2)/E*B^{0.24}/0.46*I_{se}*I_{de} \qquad \text{.......(5)}$$

It has been shown by Thomas (1993) that 1/BS is a function of the relative density of the soil (RD), the width of the footing and influence factors I_{sf} and I_{df} which are respectively, shape and depth factors at "failure" such that :-

$$BS = 0.3E\text{-}03*(RD)^{2.2}*(1+D/B)/B^{0.5}/I_{sf} \quad \ldots\ldots\ldots(6)$$

where D is the footing depth, I_{sf} =1-0.4*L/B, L is footing length and the relative density of the soil is determined from average CPT values (Thomas, 1993).

As no information relating stress-strain-time behaviour of the soils at the site was supplied to the participants, predictions of time for footing settlements to occur are based on the experience of carrying out numerous plate loading tests by the author. It has been observed that in dry granular soils, about 95% of settlement takes place as the load is applied. However, some tests on saturated sands and gravel have indicated that it may take several minutes for between 70% and 90% of the final settlement to take place. Significant plate movements can occur up to one hour from the time a load increment is applied.

Because of the relatively low ground water level at the site and the statement that the weather at the site has been very dry, with no rain during the summer of 1993, the author has assumed that almost instantaneous settlements have taken place as the loads have been applied.

SETTLEMENT PREDICTIONS

LOAD	FOOTING 1	FOOTING 2	FOOTING 3	FOOTING 4	FOOTING 5
	3.004m by 3.004m	1.505m by 1.492m	3.023m by 3.016m	2.489m by 2.496m	0.991m by 0.991m
Load for 25mm settlement Q25 on 30min load settlement curve (kN)	2835	450	1226	1786	374
Load for 150mm settlement Q150 on 30min load settlement curve (kN)	4847	650	1633	2927	457
Creep settlement between 1min and 30min for Q25 Δs(mm)	1.5	xxxxxxxxxxxxx	1.5	xxxxxxxxxxxxx	xxxxxxxxxxxxx
Settlement in year 2014 under Q25 (mm)	35	xxxxxxxxxxxxx	35	xxxxxxxxxxxxx	xxxxxxxxxxxxx

REFERENCES

1. Bjerrum,L. & Eggestad, A, (1963), Iterpretation of loading tests on sand, Proc. European Conf. Soil Mech. Found. Eng., Wiesbaden, Vol 1, pp 199 - 203.
2. Brown, P.T. & Gibson, R.E., (1972), Surface settlement of a deep elastic stratum whose modulus increases linearly with depth, Can. Geotech. Jour., No 4, pp 467 - 476.
3. Chin, F. K.,(1983), Bilateral plate bearing tests, Proc. Int. Symp. on In-Situ Testing, Paris, Vol 2, pp. 37 - 41.
4. Fox, E.N, (1948), The mean elastic settlement of a uniformly loaded area at a depth below ground surface, Proc. 2nd. Int. Conf.Soil Mech. Found. Eng., Vol 2, pp 129 - 132.
5. Giroud,J.P., (1968), Settlement of a linearly loaded rectangular area, Journ. Soil Mech. Found. Eng., ASCE., Vol 94, SM 4, pp.813 - 831.
6. Kondner, R. L.& Zelasco,J. S. (1963), A hyperbolic stress - strain formulation for sands, Proc. 2nd. Pan. Am. Conf. Soil Mech. Found. Eng., Brazil, Vol 1, pp. 289 - 324.
7. Parry, R.H.G,(1978), Estimating foundation settlements in sand from plate bearing tests, Geotechnique, Vol 28, No 1,pp 107 - 118.
8. Schleicher,F, (1926), Zur theorie des baugrundes, Bauingenieur, Vol 7,pp 931 - 935,949 - 952.
9. Thomas, D, (1964), Depsounding tests and their interpretation, Proc. Midl. Soil Mech. Found. Eng. Soc., paper 34.
10 Thomas, D, (1965), Static penetration tests in London clay, Geotechnique, Vol 15, No 2, pp. 174 - 179.
11. Thomas, D, (1966), An experimental investigation of some factors influencing penetration resistance in fine granular soils, M.Sc. thesis, University of Manchester, Institute of Science and Technology.
12. Thomas, D, (1968), Deepsounding test results and the settlement of spread footings on normally consolidated sands, Geotechnique, Vol 18, No 4, pp. 472 - 488.
13. Thomas, D, (1993), Settlement prediction of spread footings on granular soils from cone penetration tests, Research Report CE15, Dept. of Civil Engineering and Building, University of Central Queensland, Rockhampton, Australia.

ESTIMATION OF SETTLEMENTS OF SHALLOW FOUNDATIONS

M. Surendra[1], M. ASCE

ABSTRACT: This paper describes the procedures used in estimating the settlement of the five shallow foundations installed at the Texas A&M University campus as part of the "Settlement '94 Prediction Event" to be held in June 1994. The settlement of these five foundations were estimated utilizing the subsurface data from Standard Penetration Tests (SPT), Cone Penetration Tests (CPT) and Dilatometer Tests (DMT) conducted at the site.

INTRODUCTION

The settlement of the five shallow foundations were estimated using the available published procedures and the subsurface strength and compressibility characteristics of the soils at the site. The settlements were first estimated using Schmertmann's procedure (1,2) and the CPT data. These estimated magnitudes were verified using the SPT data and weighted N-values for the different strata. The preconsolidation pressure of the various strata were investigated using the DMT data. The final predicted settlements were computed using the DMT data.

SUBSURFACE CONDITIONS

The soils at the site are predominantly granular in nature. The near surface granular soils appear to have derived from alluvial deposits (3). There has been some grading that has been performed in the past. Based on the DMT data it appears that significant amount of surface compaction has been performed at the site. Based on the available data, the soil conditions as indicated by CPT-7 shows very loose soils at a depth of about 3 m. It would have been prudent to perform additional CPT soundings in the vicinity of

---

[1] Senior Project Engineer, Earth Engineering & Sciences, Inc., 3401 Carlins Park Drive, Baltimore, Maryland 21215.

CPT-7 to investigate the loose zone of soils.

The borings indicated Silty SAND to a depth of about 11 m. The standard penetration N-values ranged from 14 to 30 blows per 300 mm (blows per foot). Hard CLAY was encountered below the sand stratum and was noted to extend to maximum depths explored (about 33 m). Groundwater was noted at a depth of about 4.9 m.

ESTIMATION OF SETTLEMENTS

The settlement of the foundations were estimated based on the following available procedures.

1. CPT data: The relationship of tip resistance and elastic modulus was used in estimating settlement of the sands. The procedure suggested by Schmertmann (1,2) was used in calculating settlement of the various layers. The CPT data was also used in estimating the long term settlements.

2. SPT data: The SPT data was used in verifying the CPT data. Due to the presence of gravel, the CPT data indicated significantly high tip resistances in many cases below a depth of about 3 m. Weighted average of SPT data was also used in estimating settlement of the foundations by using Terzaghi & Peck's procedure (4). The SPT N-values were averaged based on Schmertmann's 2B-0.6 triangle and the data from 1 foot (300 mm) square footing was used to estimate the settlement of the footings.

3. DMT data: The DMT data provided during this investigation was used in the computer program ("DILLY4") provided in FHWA report, dated March 1988 (5). This data indicated significant overconsolidation of the sand stratum. However, as details regarding geologic history and compaction induced pre-stress was not available, this data was used with caution. The data indicated by DMT-1 appeared reasonable. This data was used in the procedure suggested by Schmertmann (6).

4. Pressuremeter data: The pressuremeter data was used in estimating creep effects of the sands on the settlement (7).

RESULTS

The parameters used in estimating settlement of the different foundations are summarized in Table 1. The predicted settlements are summarized on Table 2.

Table 1: Summary of Design Parameters

FOOTING	Dimensions	Boring	CPT	DMT
Footing 1	3 m x 3m D_f = 0.762 m	SPT-3 N avg = 20 Q_{25} = 2960 kN	CPT-5 qc = 90 tsf Q_{25} = 2440 kN	DMT-1 M = 700 to 1000 Bars Q_{25} = 4780 kN
Footing 2	1.5 m x 1.5m D_f = 0.762 m	SPT-4 N avg = 15 Q_{25} = 590 kN	CPT-6 qc = 53 to 56 tsf Q_{25} = 750 kN	DMT-1 M = 700 to 1000 Bars Q_{25} = 2590 kN
Footing 3	3 m x 3m D_f = 0.889 m	SPT-1 N avg = 25 Q_{25} = 3690 kN	CPT-7 qc = 33 to 100 tsf Q_{25} = 2790 kN	DMT-1 M = 700 to 1000 Bars Q_{25} = 4780 kN
Footing 4	2 m x 2m D_f = 0.762 m	SPT-5 N avg = 20 Q_{25} = 1120 kN	CPT-2 qc = 41 to 98 tsf Q_{25} = 1150 kN	DMT-1 M = 700 to 1000 Bars Q_{25} = 4020 kN
Footing 5	1 m x 1m D_f = 0.711 m	SPT-6 N avg = 18 Q_{25} = 315 kN	CPT-1 qc = 56 to 82 tsf Q_{25} = 780 kN	DMT-1 M = 700 to 1000 Bars Q_{25} = 1130 kN

D_f = Depth of Embedment 1 tsf = 107.3 kPa 1 Bar = 0.932 tsf = 100 kPa

Table 2: Estimated Settlements

	Footing 1 3 m x 3 m	Footing 2 1.5 m x 1.5 m	Footing 3 3 m x 3 m	Footing 4 2 m x 2 m	Footing 5 1 m x 1 m
Q_{25}, kN	4,780	2,590	4,780	4,020	800*
Q_{150}, kN	13,960*	2,640*	13,960*	4,690*	800*
Creep, mm	3		3		3
Settlement, in Year 2014 under Q_{25}, mm	36.5		36.5		x

Q_{25} = Load for 25 mm of settlement on the 30 minute Load Settlement Curve
Q_{150} = Load for 150 mm of settlement on the 30 minute Load Settlement Curve
Creep = Creep settlement between 1 minute and 30 minute under Q_{25}
* - Load estimated to cause Bearing Capacity failure
x = Could not be estimated

It should be noted that the settlement estimated herein assumes that the load is placed instantaneously. Depending on the loading sequence, some variations in the estimated values are expected.

The bearing capacity of Footing 5 was estimated to be 800 kN. Under this load the footing is expected to continue to settle under this load (i.e., further load increment can not be applied).

REFERENCES

1. Schmertmann, J. H., (1970), "Static Cone to Compute Static Settlement Over Sand," Journal of the Soil Mechanics and Foundation Division, ASCE, Vol 96, No. SM3, May 1970, pp. 1011-1043.

2. Schmertmann, J. H., J. P. Hartman, and P. R. Brown, (1978), "Improved Strain Influence Factor Diagrams," Journal of the Geotechnical Engineering Division, ASCE, Vol 104, No. GT8, August 1978, pp. 1131-1135.

3. Geological Highway Map of Texas, Map No.7, U.S. Geological Highway Map Series, Published by American Association of petroleum geologists, P. O. Box 979, Tulsa, Oklahoma 74101.

4. Terzaghi, K. and R. B. Peck (1967), "Soil Mechanics in Engineering Practice", John Wiley & Sons, Inc., Second Edition, 729 pp.

5. Guidelines for using the CPT, CPTU, and Marchetti DMT for Geotechnical Design, FHWA report no. PA-025+84-24, March 1988, Volume IV.

6. Schmertmann, J. (1986), "Dilatometer to Compute Foundation Settlement", Proc. Insitu'86 ASCE Specialty Conference on USE OF INSITU TESTS IN GEOTECHNICAL ENGINEERING, Virginia Tech, Blacksburg, VA, June 23-25, ASCE Geotechnical Special Publication No. 6, pp. 303-321.

7. Briaud, J.L., "Pressuremeter and Foundation Design" Proc. Insitu'86 ASCE Specialty Conference on USE OF INSITU TESTS IN GEOTECHNICAL ENGINEERING, Virginia Tech, Blacksburg, VA, June 23-25, ASCE Geotechnical Special Publication No. 6, pp. 74-115.

SPREAD FOOTING PREDICTIONS

By Kuo-Hsia Chang*, Member, ASCE

I. Method Used in the Prediction

The predictions are made by using an axial symmetrical FEM program that takes into account of the volumetric, deviator and dilatory deformations, tension cracks and shear yielding of the soils. This analysis is based on the data provided by Dr. Jean-Louis Briaud of the Texas A&M University. Detailed description of the method may be found in reference (3) at the end of this paper.

II. Results of the Prediction

Results of the prediction are summarized in the following table and curves shown on the next page.

RESULTS OF PREDICTION

No. & Size of Footing	1 3mx3m	2 1.5mx1.5m	3 3mx3m	4 2.5mx2.5m	5 1mx1m
Q25 (kN)	2370	320	1400	1000	150
Q150 (kN)	5000	500	3600	2600	450
DS (mm)	1.5	-	2.5	-	-
S20 (mm)	3.0	-	6.0	-	-

* Staff Consultant, Shannon & Wilson, Inc., 400 N.34th St. Seattle, WA 98103

III. References

Zhang, G.X., Zhang, N.R., & Zhang, F.L.(1981). "Nonlinear differential settlement analysis," Proc. XICSMFE, (2), 291-194, Stockholm.

Zhang, G.X., Zhang, N.R., & Zhang, F.L.(1984). "A comparison of some predicted and observed settlement behaviours of box-type foundations in Beijing," Proc. 3rd International Conference on Tall Buildings, Hong Kong and Guan Zhou.

Zhang, G.X., Zhang, N.R., & Zhang, F.L.(1989). "Analysis of PLT and CPT by oedo-triaxial model," Proc. XIIICSMFE, Rio de Janeiro, Brazil.

PREDICTED LOAD-SETTLEMENT CURVES FOR FOOTING #1, #3, AND #4

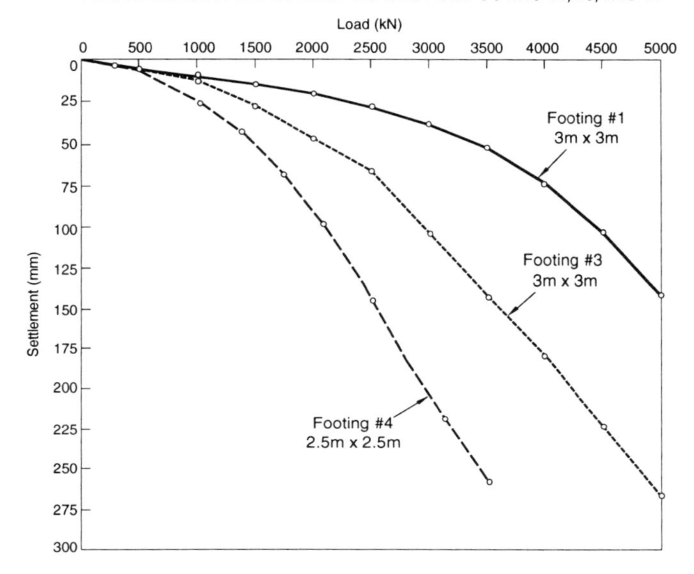

PREDICTED LOAD-SETTLEMENT CURVES FOR FOOTING #2 AND #5

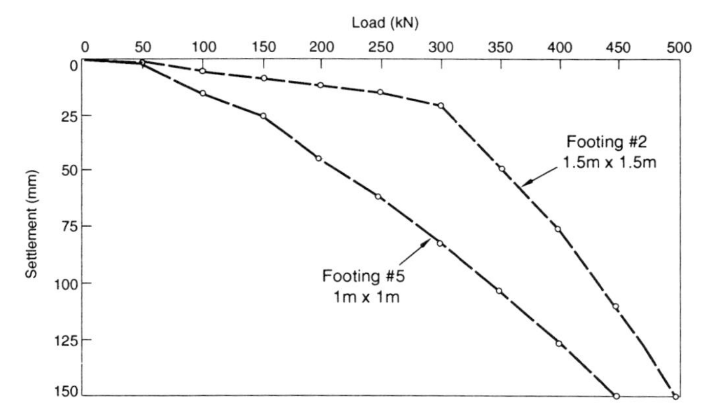

SETTLEMENT OF SPREAD FOOTINGS ON GRANULAR SOIL

By Chandra S. Brahma, FASCE[1]

ABSTRACT

The investigation for the proposed load tests on spread footings, sponsored by the Federal Highway Administration, at the National Geotechnical Experimentation Site located on the Texas A&M University, Riverside Campus reveals that predominantly sand extends from the ground surface to a depth of approximately 11 m. Below the sand and extending to a depth of about 33 m lies a very hard clay stratum. Five square footings, ranging in size from 1 m to 3 m, are scheduled to be load-tested. The paper describes the subsurface conditions, and predicted load as well as settlement of each footing, based on the results of analysis of reported laboratory as well as "in-situ" tests, and discusses prediction techniques therewith.

INTRODUCTION

Ordinarily the behavior of spread footings founded on predominantly granular soils has been estimated from either the blow count of the standard penetration tests or the cone penetration resistance. Since at least one standard penetration test boring and one piezocone penetration test sounding were performed at the approximate locations of the footings, the field tests will afford a unique opportunity to compare the accuracy of predictions, based on methods which utilize data from such field tests.

SITE CONDITIONS

The upper soil layers, approximately 11 m in thickness, consist of granular material composed of fine dense sand mixed with varying amounts of silt, clay and gravel, followed by a very hard plastic clay layer which extends to a depth of about 33 m below the finished surface. At the site, an overburden ranging in depth from 0.5 m to 1.5 m was removed prior to developing the finished surface. Depth to groundwater is 4.9 m below the finished grade. One 3 m X 3m footing, one 2.5 m X 2.5 m footing, and one 1.5 m X 1.5 m footing are located at a depth of 0.762 m below the surface. The depths to the remaining 3 m X 3 m and

[1] Professor of Civil Engineering, California State University - Fresno, Fresno, California

footing and 1 m X 1 m footing are 0.889 m and 0.711 m respectively below the finished grade..

LOAD - SETTLEMENT PREDICTIONS

The availability of results of standard penetration tests and cone penetration tests performed at the approximate locations of the footings made the prediction of loads and settlements possible by several independent methods. The load-settlement predictions reported hereinafter are based on the assumption that the subsoil profiles as depicted by the boring logs and CPT soundings are representative of the subsurface conditions at the load test site.

Empirical correlations (Terzaghi and Peck, 1967; Meyerhoff, 1956 and 1965) between standard penetration resistance, width of footing and the bearing pressure limiting specified settlement are used to predict load and settlement of individual footings. With these correlations, the standard penetration resistance (blow count N) is standardized to an energy ratio (Schmertmann, J.H., 1979) of 60%. Furthermore, the influence of effective overburden pressure was considered in the Terzaghi and Peck method.

Recently a relationship was established (Burland and Burbidge, 1985) between the compressibility of the granular soil, the width of footing and the average value of standard penetration resistance over the depth of influence of the footing. The N values are not corrected for effective overburden pressure. Burland and Burbidge, based on uncorrected blow counts, introduced the concept of compressibility being a function of density index and overburden pressure in order to determine the average settlement at the end of construction. The creep deformations for various time periods beyond the initial loading are ascertained using a factor f_t as outlined by Burland and Burbidge.

For piezocone penetration tests, the correlations between cone penetration resistance and the compessibility of the sand in the Buisman-DeBeer method (Meyerhoff, 1965) is used in estimating the settlement under a specified pressure. With this method the q_c versus depth z plot below footing is divided into a number of layers of varying thickness. For each layer the value of cone penetration resistance q_c is assumed constant. The division of q_c versus depth z plot is terminated at the depth at which the stress increment ($\Delta\sigma$) becomes less than 10% of the effective overburden pressure (σ_0') at the center of the layer. Using a constant of compressibility from a value of cone penetration resistance, the settlement of each layer is calculated for each footing.

A simplified distribution of vertical strain influence factor under the center of a shallow footing (Schmertmann et al, 1978; Schmertmann, 1970) is also used in estimating short- and long-term settlements under a specified pressure. The measured q_c versus depth z profile extending below the footing to a depth of two times the width of the footing is divided into suitable layers within each of which the value of q_c is assumed constant. Young's modulus for each layer is assumed to be 2.5 times its cone penetration resistance q_c. Based on the value of strain influence factor at the center of each layer, creep factor and strain relief factor, the settlement of each footing carrying a specified pressure is ascertained. Creep factors are used in calculating long-term settlements for footings #1 and #3.

Four pressuremeter borings, PMT-1 through PMT-4, were performed in the vicinity of footing 1. While borings PMT-1 and PMT-2 extend to a depth of 11 m below the finished surface, the remaining borings extend only to a depth of 3 m. In preborings PMT-1 and PMT-2, the pressure near the end of the linear part of the pressuremeter test curve was held for 10 minutes in order to determine effects of pressure duration on the modulus. Load-settlement predictions with pressuremeter tests are based on the consolidation (E_c) and deviatoric (E_d) modulii, using semi-empirical formulas (Menard and Rousseau, 1962; Briaud, 1986; Baguelin et al, 1978). The consolidation and deviatoric modulii are calculated for soils extending to a depth of respectively 0.5 and 8 times the footing width below the footing.

The predicted loads and settlements for footings #1 through #5 are recorded in Table 1. A reasonably good agreement exists between the loads as well as settlements calculated with the foregoing methods. For instance, the load calculated for 25 mm of settlement in 30 minutes on footing #1 is 3,250 KN with Burland and Burbidge, 2,975 KN with Terzaghi and Peck, and 3,000 KN with Meyerhoff, using the standard penetration test results. Based on the results of piezocone penetration tests, the load is 3,025 KN following the procedures outlined by Schmertmann and 3,350 KN by Buisman and DeBeer. When the results of pressuremeter tests are used, the calculated load for the same footing is 3,300 KN with the Menard method.

TABLE 1
Prediction Summary

Footing Number	Load for 25 mm of settlement in 30 min. Q_{25}(KN)	Load for 150 mm of settlement in 30 min. Q_{150}(KN)	Creep settlement between 1 min and 30 min for Q_{25} Δs(mm)	Settlement in the year 2014 under Q_{25} (mm)
1	3,025	15,200	2.4	31.5
2	775	1,990		
3	1,550	8,450	2.8	32.0
4	1,850	9,630		
5	720	1,100		

APPENDIX 1. REFERENCES

Baguelin, F., Jezequel, J.F. and Shields, D.H., The Pressuremeter and Foundation Engineering, Trans Tech Publications, Clausthal-Zellerfeld, W. Germany, 1978

Briaud, J.L., "Pressuremeter and Foundation Design", American Society of Civil Engineers - Geotechnical Publication No. 6, Virginia Polytechnic Institute, 1986

Burland, J.B. and Burbidge, M.C., "Settlement of Foundations on Sand and gravel", Proceedings, Institution of Civil Engineers, Part 1, 78, 1985

Menard, L. and Rousseau, J.L., "L'Evaluation des Tassements-Tendances Nouvelles", Sols-Soils, Volume 1, Number 1, 1962

Meyerhoff, G.G, "Penetration Tests and Bearing Capacity of Cohesionless Soils", Proceedings, American Society of Civil Engineers, 82, SM1, 1956

Meyerhoff, G.G, "Shallow Foundations", Proceedings, American Society of Civil Engineers, 91, SM2, 1965

Terzaghi, K. and Peck, R.B., Soil Mechanics in Engineering Practice, John, Wiley and Sons, New York, 1967

Schmertmann, J.H.,Hartman, J.P. and Brown, P.R, "Improved Strain Influence Factor Diagram", Proceedings, American Society of Civil Engineers, 104, GT 8, 1978

Schmertmann, J.H., "Statics of SPT", Proceedings, Journal of Geotechnical Engineering Division, American Society of Civil Engineers, GT 5, 1979

Schmertmann, J.H., "Static Cone to Compute Static Settlement over Sand", Proceedings, American Society of Civil Engineers, 96, SM 3, 1970

SPREAD FOOTING SETTLEMENT PREDICTION

By Carsten H. Floess[1], Member, ASCE

INTRODUCTION

A prediction of spread footing performance for SETTLEMENT '94 is summarized herein. Numerous prediction techniques are available based on the various test methods and data presented in the prediction request. This prediction is based primarily on the Standard Penetration Test (SPT), since it is the most common method of investigating subsurface conditions.

SETTLEMENT AT SMALL DEFLECTIONS

Available techniques for calculating settlement at small deflections are generally based on empirical correlations, one-dimensional compression methods, or elastic methods. Many of the methods tend to over-predict settlement or provide an upper bound of potential settlement. This is useful in design.

A Federal Highway Administration (FHWA) study (4) evaluated several settlement prediction methods and compared computed estimates with measured settlement of bridge piers and abutments. The study concluded that the D'Appolonia method (2, 3) was the most accurate.

D'Appolonia's method is based on an elastic modulus of compressibility that was backfigured from a limited number of measurements of footing settlement. The modulus value is presented as a function of uncorrected SPT blow counts, averaged within the depth of influence, defined as the footing width. Two influence factors account for footing shape and depth to incompressible material. These influence factors were taken from charts developed by Christian and Carrier (1).

D'Appolonia's method was developed based on measurements of footings on sand deposits. Several of the case histories in the FHWA study (4) had significant layers of silt within a depth of about twice the footing width. Settlement predictions using the D'Appolonia method at sites with silt soils tended to under-predict settlement, by as much as about 20 mm. Settlement predictions using the D'Appolonia method at sites with sand subsoils achieved remarkable accuracy, generally within about 3 mm of measured values.

---

1 Geotechnical Section Manager, Clough, Harbour & Associates,
III Winners Circle, Albany, New York 12205

The writer disagrees with the use of one footing width as the zone of influence for settlement predictions. Schmertmann et al (6) demonstrate that strains occur within a zone extending to a depth approximately twice the footing width. Accordingly, SPT blow counts were averaged within a depth equal to twice the footing width for this prediction, although these values were generally about the same as the average blow count within a depth equal to one footing width. Available data in the prediction request indicate that subsoils within the larger influence zone are generally sands with only a trace to some silt, confirming that the D'Appolonia method is appropriate for this site.

D'Appolonia's method provides modulus of compressibility values for both normally loaded and preloaded sand deposits. Available data in the prediction request indicate that the sand deposits at the test site are normally consolidated, and modulus values for normally consolidated sands were used for this prediction.

Predicted settlement at small deflections varies linearly with load. It was arbitrarily assumed that this linear relationship would be valid to a load equal to one-third of the ultimate bearing capacity, corresponding to a safety factor of 3.

SETTLEMENT AT LARGE DEFLECTIONS

There is little data in the literature regarding load settlement behavior at large strains. Three modes of shear failure are described in the literature: general shear failure, local shear failure, and punching shear failure (7). General shear failure is characterized by a well defined slip surface and failure pattern, with failure occurring suddenly. Local shear and punching shear failure have a more poorly defined failure pattern, and the point of failure is difficult to establish. The type of failure depends largely on the compressibility of the sand and the depth and size of the footing. Based on the relative density of the sand at the test site and the footing geometry, it appears that failure will probably correspond to local shear failure for all footings.

The ultimate bearing capacity of the footings was estimated using the method outlined by Vesic (7). Vesic (7) indicates that settlement of shallow footings on sand at ultimate load ranges from about 5 to 15 percent of the footing width. The smallest settlement appears to occur under conditions of general shear, while the largest settlement occurs in a punching shear failure. Accordingly, it was assumed that ultimate failure would correspond to a settlement of 10 percent of the footing width.

The load - settlement curve for each footing was completed by sketching a smooth curve between the ultimate bearing capacity at a settlement equal to 10 percent of the footing width to the end of the elastic portion of the load settlement relationship at a load equal to one-third the ultimate bearing capacity.

TIME DEPENDENT SETTLEMENT

Time dependent creep deformation between the 1 minute load-settlement curve and the 30 minute load settlement curve was estimated using available pressuremeter creep exponents. Settlements of footings 1 and 3 in the year 2014 were estimated using two-thirds of the creep correction factor proposed by Schmertmann (5).

SUMMARY

A practical and simple approach was used to predict the load-settlement curves for the test footings. The predictions are anticipated to be accurate at small deformations, with less accuracy at larger deformations. Predicted loads are summarized below:

	Footing 1 3m X 3m	Footing 2 1.5m X 1.5m	Footing 3 3m X 3m	Footing 4 2.5m X 2.5m	Footing 5 1m X 1m
Q25 @ 30 minutes	5,600 Kn	1,500 Kn	6,400 Kn	4,300 Kn	700 Kn
Q150 @ 30 minutes	11,500 Kn	2,400 Kn	11,600 Kn	7,700 Kn	1,000 Kn
Creep @ 30 minutes	1 mm		1 mm		
Q25 Settlement in 2014	33 mm		33 mm		

REFERENCES

1. Christian, J.T., and Carrier, W.D., "Janbu, Bjerrum and Kjaernsli's Chart Reinterpreted," Canadian Geotechnical Journal, Vol. 15, 1978.

2. D'Appolonia, D.J., D'Appolonia, E., and Brissette, R.F., "Settlement of Spread Footings on Sand", Journal of the Soil Mechanics and Foundations Division, ASCE, Vol. 94, No. SM3, May 1968, pp 735-760.

3. D'Appolonia, D.J., D'Appolonia, E., and Brissette, R.F., "Settlement of Spread Footings on Sand" (Closure), Journal of the Soil Mechanics and Foundations Division, ASCE, Vol. 96 No. SM2, March 1970, pp. 754-762.

4. Federal Highway Administration, "Spread Footings for Highway Bridges", Publication No. FHWA/RD-86/185, October 1987.

5. Schmertmann, J.H., "Static Cone to Compute Static Settlement Over Sand", Journal of the Soil Mechanics and Foundations Division, ASCE, Vol. 96, No. SM3, May 1970, pp 1011-1042.

6. Schmertmann, J.H., Hartman, J.P., and Brown, P.R., "Improved Strain Influence Factor Diagrams", Journal of the Geotechnical Engineering Division, ASCE, Vol. 104, No GT8, August 1978, pp 1131-1135.

7. Vesic, A.S., "Analysis of Ultimate Loads of Shallow Foundations", Journal of the Soil Mechanics and Foundations Division, ASCE, Vol. 99, No. SM1, January 1973, pp 45-73.

PREDICTIONS OF SHALLOW FOUNDATION BEHAVIOR FOR SETTLEMENT '94

Storer J. Boone[1], M. ASCE

METHODOLOGY

Six methods were used to predict foundation settlement and were chosen based on these criteria: 1) their general acceptance in foundation engineering practice; 2) use of test data in computations with little indirect correlation; and 3) overall ease in implementation. The latter factor was considered important since in practice sites with varying soil deposits and footing sizes often require many settlement predictions for a diverse spectrum of cases. Further, given inherent limitations in estimating in-situ deformation characteristics of soils, relatively simple computational techniques also allow efficient completion of parametric foundation performance studies. All techniques were implemented as suggested in the original references except as noted. The methods and data utilized were:

1. Schmertmann (1970) and Schmertmann et al. (1978): cone penetration test (CPT).
2. Buisman-DeBeer (as described by Terzaghi and Peck 1967; and Schmertmann, 1970) including Meyerhof (1965) modification: CPT.
3. NAVFAC DM-7 (1982): CPT and SPT for correlations of estimated relative density to modulus of subgrade reaction.
4. Schmertmann (1986): dilatometer (DMT).
5. Canadian Foundation Engineering Manual (1985): pressuremeter test (PMT) using initial moduli.
6. D'Appolonia et al. (1968 and 1970): Standard Penetration Test (SPT).

Stress History and Soil Profiles - Based on available data (Gibbens and Briaud 1993), the site was assumed to be normally consolidated prior to removing the 0.5 m to 1.5 m of overburden for site preparation. Soil profiles for each footing location were developed based on the nearest exploration or test point and were divided into 0.61 m (2 ft.) layers for stress and settlement computations. Average CPT data were interpreted graphically for each of these layers. Initial pressuremeter modulus values were utilized in method 5 although it has been suggested that such values may be too sensitive to disturbance and unload/reload values may be more appropriate (Mair and Wood 1988).

Stress Distribution - The prototype foundations were assumed to behave as rigid structures. Therefore, rather than estimating the stress distribution beneath a single point of a flexible footing using the Boussinesq or Westergaard approaches (as described by Holtz and Kovacs 1981) or beneath the characteristic point (Canadian Foundation Engineering Manual 1985), the simplified 2:1 method (as described by Holtz and Kovacs 1981) was employed to approximate average incremental stress increases beneath the entire footing.

---

[1] Geotechnical Project Manager; Clough, Harbour & Assoc., Albany, NY 12205

Bearing Capacity - The general equation for bearing capacity suggested by Meyerhof (1963) with shape and depth factors proposed by DeBeer (1970) and Hansen (1970) respectively, was used for prediction of ultimate foundation load capacity, Q_u. Bearing capacity factors were estimated for each footing using average angles of internal friction based on triaxial and Borehole Shear Test (BST) results within a depth equal to one footing width, and tables presented by Vesic (1973).

Settlement - Foundation loads for settlement calculation were based on average applied bearing pressures in even multiple increments of 47.9 kPa (1 ksf) up to a maximum of 3,064.4 kPa (64 ksf). Summation of incremental strains was limited to within a depth of two to three times the foundation width for methods 2 and 4. Incremental stress increases below this zone represented less than approximately 10 percent of the total applied load. The sum of incremental strains below this point also accounted for less than 10 to 15 percent of the total and was ignored. Settlement was computed for each load increment, estimation method and respective footing as illustrated on Figure 1. These results were then averaged as shown on Figure 2 since the various methods may either over or under predict settlement by factors of up to two or greater (D'Appolonia et al. 1970; Schmertmann 1970; Schmertmann et al. 1978; Baguelin et al. 1986; Schmertmann 1986; FHWA 1987). Further, in the writer's experience and opinion, use of multiple prediction methods is preferable to selecting a single approach so that assumptions, parameter selection, correlations, correction factors, and other interpretation error can be checked.

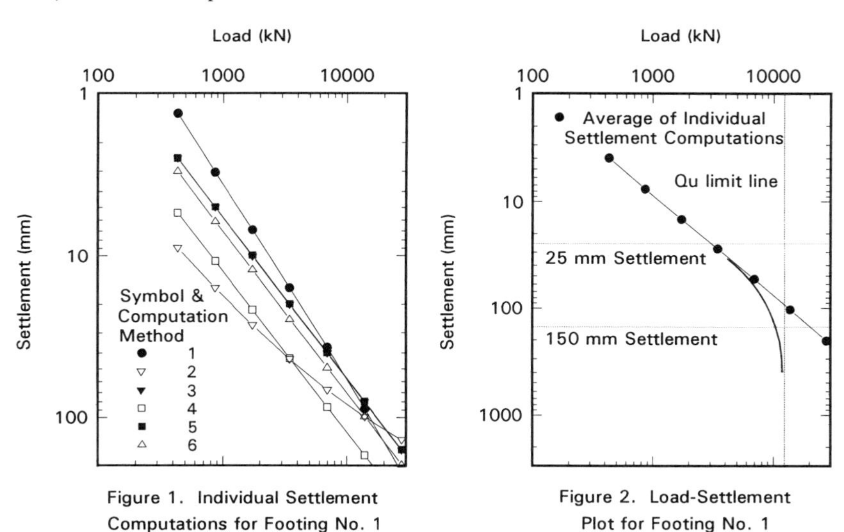

Figure 1. Individual Settlement Computations for Footing No. 1

Figure 2. Load-Settlement Plot for Footing No. 1

Time Rate of Settlement - A 30 minute period of constant applied load was used for method 1 resulting in a creep correction factor of about 0.35. The corrected settlement was considered a lower bound in light of reported conservatism of this method when the creep correction factor for long term settlement is greater than 1.0 or ignored (FHWA 1987). All other methods were considered to estimate immediate settlement within a period of 0 to 10 days (Bowles 1982) or to a maximum of 0.1 year (Schmertmann 1970). Given the case histories presented in the reviewed literature, the engineering properties of the granular soil, and the undefined time period of "immediate" settlement, the average estimated settlement was assumed to be most representative of the 30 minute load duration settlement. Based on PMT moduli versus pressure duration plots provided in an addendum to the Gibbens and Briaud (1993) data, simplified time rate of settlement estimates were made assuming the rate to be proportional to measured time dependant changes of PMT moduli. Settlement for the 1 minute load duration was then back-calculated based on the average of all creep exponents for the PMT pressure-duration tests. Using this simple, though crude, approach and a load duration of 20 years, the final settlement was estimated to be 1.36 times the initial 1 minute value.

Final Load-Settlement Behavior Prediction - One limitation of the analyses is that load-settlement relationships will be linear or nearly so as shown on Figure 1. Though appropriate for small strains of one to two percent and for settlement estimates within realistic design constraints, such linear elastic approaches are not appropriate for large strains and associated variations of deformation moduli. Therefore, to complete predictions of foundation behavior, ultimate load capacity was included on the load settlement figure and a smooth curve was fit between the average load-settlement plot and ultimate capacity limit as illustrated by Figure 2. Load settlement behavior presented by Vesic (1973) was used as a general guide for constructing the prediction figures. Foundation loads at specific values of settlement were then interpreted graphically.

PREDICTIONS & CONCLUSIONS

Table 1 provides a summary of the predicted footing loads and associated settlements.

TABLE 1	Footing 1 3m x 3m	Footing 2 1.5m x 1.5m	Footing 3 3m x 3m	Footing 4 2.5m x 2.5m	Footing 5 1m x 1m
Q25 on the 30 minute load settlement curve (kN)	3,000	1,100	1,750	2,400	600
Q150 on the 30 minute load settlement curve (kN)	10,000	2,150	10,000	7,000	760
Creep settlement between 1 minute and 30 minute of Q25 (mm)	2		2		
Settlement in the year 2014 under Q25 (mm)	32		32		

Based on the work of Vesic (1973) all footings are expected to fail in the local shear mode. Loads required to produce 25 mm of settlement (Q25) for footing 5 and 150 mm of settlement (Q150) for all footings are expected to be within the non-linear region of the load-settlement curves. Linear load-settlement behavior is expected for all other footings up to and including Q25. The lower Q25 predicted for footing 3 results from the loose sand zone indicated by CPT-7 between the depths of about 2.3 m and 3.0 m and subsequent computations based on the CPT data. Settlement at failure, expressed as a percent of foundation width, is predicted to occur between 5 percent for the smallest footing and 12 to 17 percent for the two largest footings.

Certainly more sophisticated analyses could have been performed. However, the intent of the presented work was to: 1) prepare predictions based on methods economically implemented for small and large foundation engineering projects; 2) utilize commonly available and economical in-situ testing techniques such as the SPT and CPT, and those becoming more accessible such as the PMT and DMT; 3) participate in the event as a practicing geotechnical engineer; and 4) to gage the writer's own experience and approaches to prediction of foundation behavior.

APPENDIX. REFERENCES

Baguelin, F.J., Bustamante, M.G., Frank, R.A. (1986). "The Pressuremeter for Foundations: French Experience," *Use of In-situ Tests in Geotechnical Engineering, ASCE Geotechnical Special Publication No. 6*, ASCE, 31-46.

Bowles, J.E. (1982) *Foundation Analysis and Design, Third Edition*, McGraw-Hill, Inc., New York, 185.

Canadian Foundation Engineering Manual, (1985). Canadian Geotechnical Society, Montreal.

D'Appolonia, D.J., D'Appolonia, E., Brissette, R.F. (1968). "Settlement of Spread footings on Sand," *Journal of the Soil Mechanics and Foundation division*, ASCE, 94(3), 735-759.

D'Appolonia, D.J., D'Appolonia, E., Brissette, R.F. (1970). "Settlement of Spread footings on Sand," (Closure), *Journal of the Soil Mechanics and Foundation Division*, ASCE, 96(2), 754-762.

DeBeer, E.E. (1970). "Experimental Determination of the Shape Factors and Bearing Capacity Factors of Sand," *Geotechnique*, 20(4), 387-411.

FHWA Federal Highway Administration (1987). *Spread Footings for Highway Bridges FHWA/RD-86/185*, Federal Highway Administration, Washington, DC.

Gibbens, R., Briaud, J.L. (1993). "Data and Prediction Request for the Spread footing Prediction Event Sponsored by FHWA at the Occasion of the ASCE Specialty Conference: Settlement '94," Texas A&M University, TX.

Hansen, J.B. (1970). "A Revised and Extended Formula for Bearing Capacity," *Bulletin 28*, Danish Geotechnical Institute.

Holtz, R.D., Kovacs, W. D. (1981). *An Introduction to Geotechnical Engineering*, Prentice-Hall, Inc., NJ.

Mair, R.J., Wood, D.M. (1987). *Pressuremeter Testing Methods and Interpretation*, Construction Industry Research and Information Association, Butterworths, London.

Meyerhof, G.G. (1963). "Some Recent Research on the Bearing Capacity of Foundations," *Canadian Geotechnical Journal*, 1(1), 16-26.

Meyerhof, G.G. (1965). "Shallow Foundations," *Journal of the Soil Mechanics and Foundations Division*, ASCE, 82(1), 21-31.

NAVFAC DM7.1, (1982). *Soil Mechanics, Design Manual 7.1*, Department of the Navy, Washington, DC.

Schmertmann, J.H. (1970). "Static Cone to Compute Static Settlement Over Sand," *Journal of the Soil Mechanics and Foundation Division*, ASCE, 96(3), 1011-1043.

Schmertmann, J.H., Hartman, J.H., Brown, P.B. (1978). "Improved Strain Influence Factor Diagrams," *Journal of Geotechnical Engineering*, ASCE, 104(8), 1131-1135.

Schmertmann, J.H. (1986). "Dilatometer to Compute Foundation Settlement," *Use of In-situ Tests in Geotechnical Engineering, ASCE Special Publication No. 6*, ASCE, 303-321.

Terzaghi, K., Peck, R. B. (1967). *Soil Mechanics in Engineering Practice, Second Edition*, John Wiley & Sons, NY, 517-518.

Vesic, A.S. (1973). "Analysis of Ultimate Loads of Shallow Foundations," *Journal of the Soil Mechanics and Foundations Division*, ASCE, 99(1), 45-73.

SPREAD FOOTING SETTLEMENT PREDICTION FOR THE NATIONAL GEOTECHNICAL EXPERIMENTATION SITE PREDICTION EVENT: SETTLEMENT '94

P. Cooksey[1], Q. Tang[1], M. Zhao[1], M. Mauldon[2], E. Drumm[3]

INTRODUCTION

This report provides the load-settlement predictions of five shallow spread footings at the National Geotechnical Experimentation site located on the Texas A&M University campus. The prediction includes parameter estimation and load-displacement analysis. The parameter estimation was based on the field and laboratory test data provided by Gibbens and Briaud (1993). The settlement predictions were developed by combining results from three separate analyses. The details of parameter estimation and analysis are discussed in the following text.

SITE CHARACTERIZATION

According to Gibbens and Briaud(1993), the soil at the experimentation site is predominantly medium dense silty fine sand to a depth of 11 meters underlayed by 22 meters of very hard plastic clay. The water table was observed at a depth of 4.9 meters. The soil properties were given by 13 sets of field investigation and laboratory test data. Young's modulus was calculated using results from the Standard Penetration Test (SPT), PiezoCone Penetration Test (CPT) and Dilatometer Test (DMT). The Young's modulus profile beneath each footing was determined by contouring the values obtained at one meter depth increments. A typical contour map, created by SURFER (Golden Software, 1993), is shown in Figure 1. Poisson's ratio was determining using the friction angle from the Borehole Shear Tests.

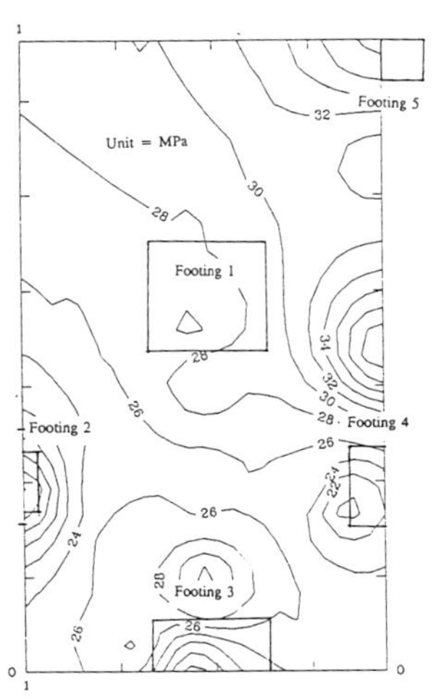

Figure 1
Typical contour map of Young's modulus.

[1]Graduate Student, [2]Assistant Professor, [3]Associate Professor. Institute for Geotechnology, Dept. of Civil and Environmental Engineering, The University of Tennessee, Knoxville.

SETTLEMENT CALCULATIONS

Settlement calculations were performed using both conventional and numerical methods. The three analyses performed are discussed below. When feasible, the same values for all parameters (i.e. Young's modulus and Poisson's ratio) were used in each method. The load corresponding to 25 mm of settlement from the 30 minute load-settlement curve, Q_{25}, was calculated using each method. The load corresponding to 150 mm of settlement, Q_{150}, was determined only when 150 mm did not exceed the settlement of the footing at bearing capacity. The bearing capacity was determined using Meyerhof's (1963) bearing capacity factors.

Method 1

Method 1 calculated the settlement using the computer program CSANDSET (Knowles, 1991). CSANDSET calculates the settlement of shallow footings using fifteen different methods. Each method calculates a Young's modulus or necessary parameter based on the SPT and CPT data input. The resulting settlements varied by as much as a factor of 15 for a given load. For Q_{25}, the immediate load settlement curves for each of the footings were based on an average of the four methods that fell between the median and mean results. In all cases the four methods were Elastic Theory, Schmertmann (1970), Bowles (1977) and NAVFAC DM 7.1 (Department of the Navy, 1982). The Q_{150} values were determined using Schmertmann (1978); this method provided larger settlements for a given load without exceeding the bearing capacity of the soil.

Method 2

Method 2 was based on the procedure given by Leonards and Frost (1988) which accounts for the reduced compressibility of overconsolidated granular material. The DMT and CPT data were used directly in the calculations. As expected, when compared to elastic methods this procedure results in reduced settlements for any given load. Q_{150} could not be determined from this method, as that load exceeds the bearing capacity of the soil.

Method 3

The finite element method was also used to predict the behavior of shallow footings. The ABAQUS code (HKS, 1992) was used for this analysis. The footing and soil were simulated by 8-node axisymmetric solid elements, with the square footing converted to an equivalent circular footing. The footing was represented by a linear elastic material model and the soil by an elastic-perfect plastic Drucker-Prager model. Yield was assumed to be governed by the Mohr-Coulomb surface with a friction angle, $\phi = 33°$. the soil above the footing base was assumed to have $c = 50$ kPa and the soil below $c = 0.05$ kPa. The displacement control loading method with geostatic stress was used to obtain the vertical displacement as a function of load. Non-associated flow rule was adopted in the calculation.

METHOD COMPARISON

Results from methods 1 and 2 were compared with Schmertmann's (1978) method using the Young's Modulus from the contour plots. The resulting load for 25 mm of settlement was found to be less than the values from Method 1 for all footings except #3. The Q_{25} from Method 2 exceeded the values from Schmertmann's method for all five footings.

Load-displacement relationships for Footing 2 from the numerical analysis, Leonards & Frost (1988) method and Schmertmann (1978) method are shown in Figure 2. The limiting pressures calculated using Meyerhof (1963) and Terzaghi(1943) bearing capacity factors are represented by the dashed lines. The linear elastic response is also shown. The calculated values of Q_{25} and Q_{150} for each footing from all three methods are presented below in Table 1.

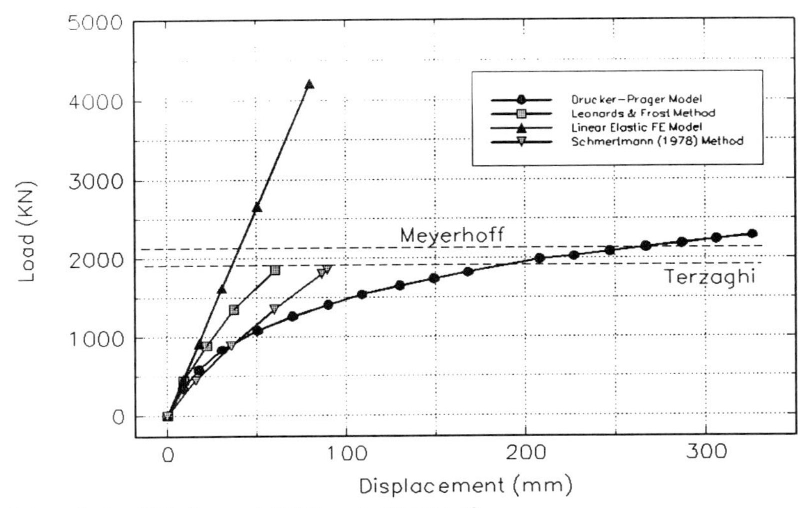

Figure 2, Load-Displacement Relations for Footing #2

		Footing 1	Footing 2	Footing 3	Footing 4	Footing 5
Q_{25} (kN)	Method 1	3005	713	1900	1721	546
	2	4300	1000	4250	2300	515
	3	2590	860	2620	1870	630
Q_{150} (kN)	Method 1	8360	2000	3468	5500	---
	2	---	---	---	---	---
	3	8420	1980	8480	5890	920

Table 1 Preliminary predictions from three separate analysis. Method 1=CSANDSET, Method 2=Leonards and Frost, Method 3=Numerical.

The settlement occurring between 1 minute and 30 minute is assumed to be caused by creep. The 30-minute settlement values were determined by extending the pressuremeter holding test results provided by Gibbens and Briaud (1993). The resulting Young's modulus for 30 minutes is 95% of the 1 minute modulus, or $E_{30}/E_1 = 0.95$.

The prediction for settlement in the year 2014 for footings 1 and 3 was calculated in two ways. First, a further extension of the method described above was used. It was found that no significant decrease in Young's modulus occurrs after 100 days. At 100 days the modulus decreased by a factor of 0.84. To account for secondary creep the modulus was further reduced for a final modulus of $0.80{*}E_{1min}$. This results in a 20 year settlement 1.25 times greater than the calculated immediate settlement. For comparison, Schmertmann's correction (1970), C_2. was calculated. For 20 years, the resulting settlement would be 1.46 times greater than the calculated immediate settlement. The final prediction value relied primarily on the 0.8 reduction in Young's modulus.

PREDICTION

The final predictions, Table 2, were obtained by examining the range of results from the three methods discussed above and determining a value which best represented the engineering judgement of the first three authors. The prediction for Q_{25} is based primarily on the conventional calculation results, Methods 1 and 2. The load-displacement relationship for small displacements is assumed to be elastic. The Q_{150} predicted values are derived from the numerical analysis since the conventional methods could not be used reliably at such large loads.

	Footing 1	Footing 2	Footing 3	Footing 4	Footing 5
Q_{25} (kN)	3700	900	3750	2100	550
Q_{150} (kN)	9000	2000	9000	6400	900
ΔS (mm)	1.3	---	1.3	---	---
S_{2014} (mm)	30	---	30	---	---

Table 2. Final settlement predictions.

REFERENCES

Bowles, J. E. (1977), *Foundation Analysis and Design*, 2nd ed., McGraw-Hill, New York.

Department of the Navy, NAVFAC (1982), *Soil Mechanics, Design Manual 7.1*, Alexandria, VA.

Gibbens, R. and Briaud, J.,(1993) *Data and Prediction Request for the Spread Footing Prediction Event*.

Golden Software (1993), *SURFER Reference Manual*, Version 4.

Hibbitt, Karlsson and Sorensen, Inc.,(1992), *ABAQUS User's Manual* and *ABAQUS Standard Example Problems*.

Knowles, V. R. (1991), *Settlement of Shallow Footings on Sand: Report and User's Guide for Computer Program CSANDSET*, Department of the Army, Technical Report ITL-91-1, June 1991.

Leonards, G. A. and Frost, J. D. (1988), Settlement of Shallow Foundations on Granular Soils, *Journal of Geotechnical Engineering*, Vol. 114, No. 7, July 1988, pp. 791-809.

Meyerhof, G.G. (1963), Some Recent Research on the Bearing Capacity of Foundations, *CGJ*, Ottawa, vol. 1, no. 1, Sept., pp 16-26.

Schmertmann, J. H. (1970), Static Cone to Compute Static Settlement Over Sand, *Journal of the Soil Mechanics and Foundations Division*, Proceedings of the American Society of Civil Engineers, Vol. 96, No. SM3, May 1970, pp. 1011-1043.

Schmertmann, J. H. (1978), *Guidelines for Cone Penetration Test: Performance and Design*, FHWA-TS-78-209 (report), U.S. Dept. of Transportation, 145 pp.

Terzaghi, K. (1943) *Theoretical Soil Mechanics*, John Wiley and Sons, N.Y. 510 pp.

Figure 1 Empiricism in the World of Soil Mechanics

Prediction of Foundation Load/Settlement Characteristics for Settlement '94

by John Scott and Keith Nicholls

Prediction of Foundation Load/Settlement Characteristics

John Scott[1] and Keith Nicholls[2]

1.0 Introduction

This paper presents the authors' methodology behind the prediction of the load required to produce a specified settlement of five footings under sustained load. The data was provided in the Gibbens and Briaud (1993) report and subsequent correspondence. The assessment uses published SPT and CPT based settlement and bearing capacity prediction methods. Several settlement prediction methods were utilised and compared. The methods were, in some cases, modified to take account of the fact that this was a prediction exercise, rather than a more conservative design exercise. In addition, a simple elastic settlement approach was also adopted using in-situ test data/elastic parameter correlations. The final predictions were chosen after scrutinising the site investigation data and comparing all the results.

2.0 Soil Behaviour under Load.

Without specific knowledge of the area's geology, it has been assumed that the soils encountered at the Riverside Campus are fluvial in origin and relatively young. The soil is expected to be lightly over-consolidated, as a result of changes in effective stress due to groundwater variability and as a result of the removal of a small amount of soil across the site.

The above, together with preliminary settlement calculations, suggested that the upper 5m to 6m of the soil profile will be behaving largely in a normally consolidated manner under the loads applied to give a footing settlement of 25mm.

3.0 Design Assumptions

In order to undertake the prediction exercise, a number of assumptions were made to facilitate the calculations, namely;

(a) Interaction between the footings was not taken into account. This assumption would generally appear reasonable, except where the footings are at or near 'classical' bearing capacity failure. Any interaction effect is expected to result in underprediction of the loads calculated. It was further assumed that the installation of the reaction system did not effect the soil properties.

(b) Based on inspection of the CPT piezo element data, it was assumed that the upper (sandy) soil profile behaved in a drained manner upon loading. The strain hardening effects of the load/reload cycles were assumed to be taken into account in the soil stiffness profiles adopted.

(c) It was assumed that the load that caused 150mm of footing settlement was at, or close to, the ultimate load capacity of the footing.

(d) A rigid boundary was assumed to occur at 2B (B = footing width) or 6m depth, whichever was shallower. Consolidation of the underlying clay/silt lenses and layers was ignored. The presence of the water table was assumed to be accounted for in the CPT/SPT/Dilatometer results. The soils were considered to be elastically homogeneous, ie $E_{horiz.} = E_{vert.}$.

---

[1] Geotechnical Engineering Office, Civil Engineering Department, Hong Kong Government.

[2] Geotechnical Engineering Office, Civil Engineering Department, Hong Kong Government (now with Wardell Armstrong Consulting Engineers, UK)

4.0 Methods Adopted and Specific Assumptions

4.1 Load at 25mm Settlement

The following three methods were used to estimate the load at 25mm settlement, which was assumed to be at the end of 30min sustained load cycle.

4.1.1 Burland and Burbidge (1985)

The methodology adopted is that of Burland and Burbidge (1985). The SPT N values in the depth of influence (Z_i) below the footing level were corrected for the presence of silty sands (Terzaghi and Peck, 1948). No correction of the SPT N value was made for depth. The SPT profile used for each footing was taken from the closest available borehole. The results of this exercise are summarised in Table I below (refer to Section 4.1.4). Bearing capacity calculations for the $1m^2$ footing (Footing No. 5) revealed that the bearing pressure estimated by the Burland and Burbidge method was close to bearing capacity failure, and the estimated load was therefore not calculated.

4.1.2 Elastic Analysis

A 3-D elastic analysis was also adopted using a simple computer program, which assumes a flexible footing founded on horizontally layered, constant elastic (E) modulus strata. In this case, the layer thickness adopted ranged from 0.25m immediately below the footing to 1.0m at depth. The constant elastic modulus adopted for each layer (calculated at the centre of each layer) was taken as;

$$\mathbf{E = 13 + 5.3z\ (MPa)} \qquad \mathbf{(1)}$$

where z is the depth below ground surface level to a depth of 5m. Between 5m and 6m, a constant E = 30MPa was assumed, to account for the presence of silt and clay interbeds. This profile was developed considering both the envelope of CPT results assuming $E = 3q_c$, and from the envelope of dilatometer results assuming $E = E_D$ (both after Lunne et al, 1989). The elastic profile (Eq. 1) was assumed where the applied bearing pressure (q_{app}) was less than or equal to 70% of the calculated bearing capacity failure (q_{ult}) pressure. The elastic profile was factored by 0.8 for an $q_{app.}$ between 70% and 80% of q_{ult}. This was to account for strain softening as the footing approached q_{ult} and was based on inspection of the load/settlement behaviour of pile and plate load test data. To account for footing rigidity, the settlement (s) of each footing was assumed to be equal to;

$$\mathbf{s_{(rigid)} = 0.5[s_{(centre)} + s_{(edge)}]_{(flexible)}} \qquad \mathbf{(2)}$$

A drained Poisson's ratio of 0.25 was also assumed, and the assumptions outlined in Section 3.0 are also relevant. The results of this exercise are also summarised in Table I.

4.1.3 Schmertmann (1970,1979)

The approach of Schmertmann (1970) was also utilized using the I_z values published by Schmertmann et al (1978) and assuming the elastic modulus profile as adopted for the elastic analysis (Eq 1). The results of this exercise are also summarised Table I. The $1m^2$ footing was ignored for this calculation for the same reasons given in Section 4.1.1.

4.1.4 Summary of Calculations

Table I

Predictions of Load (kN) to Give 25mm Settlement of Footing

Method of Analysis	Footing No.				
	1	2	3	4	5
Burland and Burbidge (1985)	3350	1250	3000	2350	-
Elastic Analysis	3000	1250	3000	2375	620
Schmertmann (1970,1978)	2600	900	2600	1925	-
Authors' Predictions (Q_{25})	3350	1250	3000	2375	620

It is noted from Table I that the Burland and Burbidge (1985) method gives very similar results to the elastic analysis. The Schmertmann et al approach gives slightly lower load predictions. After reviewing the above calculations and the site investigation data, the "Authors' Predictions" (Table I) were chosen as the loads most likely to give a footing settlement of 25mm after 30mins sustained load.

4.2 Load at 150mm Settlement (Q_{150})

This load assessment was based on the assumption that 150mm of settlement represents bearing capacity failure. This is considered a better representation for the 1.0m^2 and 1.5m^2 footings, with the settlement being equivalent to 10% to 15% of the footing width. The calculations were based on ultimate bearing capacity theory, as described by Bowles (1988) (presenting the work of Vesic, 1974), using an average ϕ' determined from the triaxial testing (ie $\phi' = 36°$, which gives $N_q = 37.7$ and $N_\tau = 56.2$). The Q_{150} predictions are summarised in Table II (refer to Section 5.0 below).

4.3 Creep under 29min Sustained Load

The movement under sustained load was based on the assumption that the pressuremeter test (PMT) creep results provided a reasonable estimate of the likely creep movements. Inspection of the results indicated about a 2% to 4% reduction in stiffness per log cycle of time (between 1min. and 10mins.) for the upper soil profile. Extrapolating the log data and assuming an average of 3% reduction in stiffness with log cycle time, the authors arrived at a decrease in stiffness of about 4% over 29minutes. It was further assumed that the increase in settlement was directly proportional to the reduction in stiffness. These assumptions gave a creep movement of 1mm over 29mins sustained load (Table II).

4.4 Creep under 20 Years of Sustained Load.

Two methods were used to estimate the load under 20 years sustained load. The first was to adopt the same approach and assumptions as outlined in Section 4.3. The Q_{25} load prediction (Table I) to give 25mm settlement at the end of 30mins was used as the reference point settlement to estimate the creep after 20 years. This approach gave an estimate of 5mm creep over 20 years. Secondly, the Burland and Burbidge approach for the static load case was adopted, which gave an estimate 12mm of creep settlement during the same period. The latter number was chosen as the authors' prediction of total creep settlement over this period and was added to the Q_{25} to give a total predicted settlement after 20 years (Table II).

5.0 Summary

The authors considered it a useful exercise to use published settlement and bearing capacity prediction methods to predict the loads required to give specified settlements of footings. A summary of the predictions is presented in Table II. The exercise was enjoyable and the authors look forward with interest to the load testing of the footings for a comparison with the results and a comparison with prediction methods based on more theoretically rigorous methods of analysis.

Table II

Authors' Predictions of Foundation Load/Settlement Characteristics

Predicted Quantity	Footing No.				
	1	2	3	4	5
Load for 25mm of settlement (Q_{25}) on the 30min load settlement curve (kN)	3300	1250	3000	2380	620
Load for 150mm of settlement (Q_{150}) on the 30min load settlement curve (kN)	15800	2400	15800	9600	850
Creep settlement between 1 min and 30min for Q_{25} (mm)	1		1		
Settlement in year 2014 under Q_{25} (mm)	37		37		

6.0 Acknowledgements

This paper is published with the permission of the Director of Civil Engineering, Hong Kong Government. Thanks are due to colleagues in the Geotechnical Engineering Office who provided useful discussion and comments.

7.0 References

Bowles, J.E. (1988). Foundation Analysis and Design. 4th Ed. McGraw-Hill Book Company. New York.

Burland, J.B. and Burbidge, M.C. (1985). Settlement of foundations on sand and gravel. Proc. Inst. Civ. Engrs Vol. 78(1), pp 1325-1381.

Gibbens, R. and Briaud, J.L. (1993). Prediction Request for Settlement '94 Conference. Texas A&M University. [Plus subsequent correspondence].

Lunne, T., Lacasse, S. and Rad, N.S. General report/Discussion session 2: SPT, CPT, pressuremeter testing and recent developments in in-situ testing - Part 1: All tests except SPT. (1989). Proc. XII Int. Conf. SMFE, Rio de Janeiro, Vol. 4 pp 2339-2403.

Schmertmann, J.H. (1970). Static cone to compute static settlement over sand. JSMFD, ASCE, Vol. 96, No. SM3, pp 1011-1043.

Schmertmann, J.H., Hartman, J.P. and Brown, P.R. (1978). Improved strain influence factor diagrams, JGED, ASCE, Vol. 104, No. GT8, pp 1131-1135.

Terzaghi, K. and Peck, R.B.(1948). Soil Mechanics in Engineering Practice. John Wiley and Sons, New York

Vesic, A.S.(1974). Foundation Engineers Handbook. Van Nostrand Reinhold, New York.

UF Spread Footing Prediction

by

F.C. Townsend[1]
P. Ruesta[2]
R. Morgan[2]

INTRODUCTION

The FHWA/TA&M Settlement '94 prediction event provides a complete array of insitu and laboratory data, plus full-scale footing tests. Consequently, the event offers a unique opportunity to evaluate/compare various predictive methods. Accordingly, UF elected to use; (1) CPT, (2) DMT, and (3) FEM based predictions.

CPT - Schmertmann's (1970) method is utilized in the program SANDSET (Patev,ND) with the predictions presented in Table 1. The CPT log corresponding to each footing was used as is; except CPT 7, which exhibits a void at 3m (10ft.). For this case (Ftn # 3) we used data from boring STP-1 and a $q_c/N = 3.5$ to eliminate the void.

Traditionally, we have found the CPT method to overpredict conservatively settlements, which is confirmed by the data in Table 1.

DMT - Settlement predictions based upon DMT used Schmertmann's concepts (1986), which are enhanced by Dumas' (1990) program DILLYSET. DMT-3 data were used for Ftn #'s 1 and 4 as thrust values were provided, and data reduction of DMT readings to OCR, ϕ, and P_c are dependent upon thrust readings. DMT-1 + DMT-3 thrust values were used for Ftn's # 3 and 3. Two difficulties with the DMT approach are; (1) that frequent readings are required near the surface zone of influence, and (2) sands tend to be indicated as over-consolidated, which leads to stiffer modulus values (reduced settlements).

Our experience (Skiles, 1992) is that DILLYSET tends to unconservatively underestimate settlements. This experience is confirmed by the data in Table 1.

FEM - An axi-symmetric FEM analyses incorporating the Duncan-Chang model (1970) and an equivalent area diameter was used for predicting the settlements of Footings # 1,3,and 4. The model parameters were derived from the furnished triaxial test data as follows: $K = 26{,}220\text{kPa}$, $n = 0.87$, $R_f = 0.9$, $G = 0.4$, $F = .01$, and $d = 4.33$.

---

[1] Professor, [2] Graduate Student, Department of Civil Engineering, 345 Weil Hall, University of Florida, Gainesville, FL 32601.

DISSUSSION

The data presented in Table 1 reveal that the DMT and FEM predictions are in agreement for 25mm of settlement, while the CPT load predictions at this settlement are deficient (too soft). Conversely, at 150mm of settlement, the CPT and FEM settlements are in agreement, while the DMT load predictions are excessive (too stiff). Accordingly, "engineering judgement" is required for an accurate prediction, as reflected in Table 2. We believe the DMT and FEM predictions for 25mm are correct, while the average of the CPT and DTM results represents our estimate of the load required for 150mm of settlement.

REFERENCES

Dumas, C. E. (1990) "Dilly - User's Manual" MS Report, dept. of Civil Engineering, UF, Gainesville, FL 32611.

Duncan, J.M, and Chang, C-Y, (1970) "Nonlinear Analysis of Stress and Strain in Soils", ASCE Jrn SMFE, Sept. SM 5, 1629-1654.

Patev, R. C. (no date) "CSANDSET" USAEWES Information Technology Lab, Vicksburg, MS. 39180

Schmertmann, J.H. (1970) "Static Cone Test to Compute Static Settlement over Sand" ASCE J-SM&FE, Vol. 96, SM3, pp 1011-1043.

Schmertmann, J.H. (1986) "Dilatometer to Compute Foundation Settlement" ASCE Proc. In Situ '86 - Specialty Conf. on Use of In-Situ tests and Geotechnical Engineering, Blacksburg, VA 303-321.

Skiles, D.L. (1992) " Evaluation of Predicting Shallow Foundation Settlement in Sands from Dilatometer Tests" MS Thesis, Dept. of Civil Engr., UF, Gainesville, FL 32611.

Table 1 Summary of Loads (Kn) for Settlement Predictions

	Ftn #1 3m x 3m			Ftn #2 1.5m x 1.5m		Ftn #3 3m x 3m			Ftn #4 2m x 2m			Ftn #5 1m x 1m
	CPT	DMT	FEM	CPT	DMT	CPT	DMT	FEM	CPT	DMT	FEM	CPT
δ = 25 mm @ 30'	2284	6000	5400	461	2100	1862	4250	5400	1080	3500	3240	404
δ = 150 mm @ 30'	9670	36,000	7740	1911	12,480	7989	25,250	7740	4608	21,000	3760	1603

Table 2 UF's Estimate of Loads (Kn) Required for 25 mm and 150 mm Settlement

	Ftn #1 3m x 3m	Ftn #2 1.5m x 1.5m	Ftn #3 3m x 3m	Ftn #4 2m x 2m	Ftn #5 1m x 1m
δ = 25mm	6000	2100	5400	3500	404
δ = 150mm	22,835	7,195	16,620	12,804	1603

Spread Footing Settlement Prediction for
ASCE Specialty Conference "SETTLEMENT'94"

By Jon R. Foshee, P.E.

This paper presents the results of spread footing settlement calculations done for the "SETTLEMENT '94" prediction symposium. Load tests performed on five footings located at the National Geotechnical Experimentation Site on the Texas A&M University Riverside Campus will be used to evaluate the accuracy of settlement estimates submitted by various predictors. The footings were constructed on a predominately sand profile and extensive in-situ soil tests were done to define the site subsurface conditions.

The attached estimated settlements (Table 1) were calculated using the computer program "SETTL/G" developed by Geosoft(1988). "SETTL/G" is designed to calculate stresses and settlements under uniformly loaded rectangular areas. In calculating stresses, the affects of adjacent loaded areas are taken into account. The program has four different stress distribution options and soil compressibility can be input three different ways.

For this study, soil compressibility was defined in terms of elastic properties and a Westergaard stress distribution was used. The soil elastic modulus (E) was estimated from the Cone Penetrometer Test results and a relationship published by Robertson and Campanella(1984) which relates Cone Bearing and Drained Secant Modulus at 25% failure stress (E_{25}) for normally consolidated, uncemented quartz sands. Since the groundwater table was relatively deep at this site, the soil elastic moduli were not adjusted for overconsolidation. The Poisson's ratio used was 0.4 above the water table and 0.49 below the water table as discussed on page 28 of the June 1993 symposium test data report.

A method for evaluating creep settlement between 1 minute and 30 minutes for the Q25 load was not readily available so this estimate was not attempted. The estimate of settlement in the year 2014 under the Q25 load was determined using Schmertmann's(1970) long term creep correction factor(C_2).

REFERENCES

Geosoft(1988), SETTL/G Settlement and Stress Distribution Analysis, 1442 Lincoln Avenue, Suite 146, Orange, CA 92665.

Robertson and Campanella(1984), Guidelines For Use and Interpretation of the Electric Cone Penetration Test, The University of British Columbia, Second Edition, pp. 79-81.

Schmertmann, J. H.(1970), Static Cone to Compute Static Settlement over Sand, JSMFD, vol. 96, SM3, May, pp. 1011-1043.

District Geotechnical Engineer, Florida Department of Transportation, District 5 Materials & Research Office, 719 South Woodland Blvd., DeLand, FL 32720

TABLE 1

	Footing 1 3 m x 3 m	Footing 2 1.5m x 1.5m	Footing 3 3 m x 3 m	Footing 4 2 m x 2 m	Footing 5 1 m x 1m
Load for 25 mm of settlement Q25 on the 30 minute load settlement curve (Kn)	1,838 Kn (413 kips)	564 Kn (127 kips)	655 Kn (147 kips)	1,190 Kn (268 kips)	423 Kn (95 kips)
Load for 150 mm of settlement Q150 on the 30 minute load settlement curve (Kn)	9,016 Kn (2027 kip)	2,720 Kn (611 kips)	2,641 Kn (594 kips)	5,949 Kn (1337 kip)	2,104 Kn (473 kips)
Creep settlement between 1 minute and 30 minute for Q25, s (mm)	--		--		
Settlement in the year 2014 under Q25 (mm)	38 mm		37 mm		

Prediction of Settlement of Footings on Sand

G. Mesri[1], M. ASCE and M. Shahien[2], A.M. ASCE

INTRODUCTION

Load and settlement predictions were made for the five footings on sand described by Gibbens and Briaud (1993). Two separate and independent methods were used for settlement analysis. The square, rigid, concrete footings were placed at depth of 0.76 m on a medium dense silty fine sand that extends to a depth of 3.5 m below a graded ground surface. In preparing for the footing test program, the site was graded by removing between 0.5 to 1.5 m of surface sand. The fine sand is underlain by medium dense silty sand with clay and gravel. The water table is at 4.9 m below the graded ground surface. Subsurface conditions have been described in detail by Gibbens and Briaud (1993).

METHOD USING DSPT N (Method A)

Settlement of a footing on sand consists of an end-of-construction (EOC) settlement, s_c, and a post-construction secondary settlement, s_s. The method for computing s_c, using Drive Sampler Penetration Test (DSPT) N values, is patterned after the work of Burland and Burbidge (1985). The settlement equation for an embedded footing on normally consolidated sand is:

$$s_c = Z_I \frac{1}{3} \bar{m}_v \sigma'_{vo} + Z_I \bar{m}_v (q - \sigma'_{vo}) \tag{1}$$

where Z_I is the depth of influence in the granular soil beneath the footing beyond which vertical strains are negligible, $\bar{m}_v$ is the average coefficient of vertical compression of granular soil thickness Z_I, q is the average gross bearing pressure on the footing, defined as q = P/A, where P is the total load on the footing and A is the area of the footing, and σ'_{vo} is the pre-construction effective vertical stress at the level of the base of the footing. The following empirical expressions, based on settlement of

[1]Professor of Civil Engineering, University of Illinois at Urbana-Champaign, Urbana, Illinois 61801

[2]Graduate Student, University of Illinois at Urbana-Champaign, Urbana, Illinois 61801

foundations, tanks, and embankments on sands and gravels, are used to determine Z_I in meters and $\bar{m}_v$ in MPa^{-1}:

$$Z_I = B^{0.75} \tag{2}$$

$$\bar{m}_v = \frac{1.7}{\bar{N}^{1.4}} \tag{3}$$

where B is the breadth of foundation in meters, and $\bar{N}$ is the arithmatic mean of the measured DSPT N values over thickness Z_I defined by Eq. 2.

It is assumed that primary consolidation of the granular soil is completed by the end of construction, and only secondary compression follows thereafter. It is also assumed that the final construction day represents the duration of the last primary compression increment. Thus, the following equation is used to compute secondary settlement:

$$s_s = \varepsilon_\alpha Z_I \log \frac{t}{1 \text{ day}} \tag{4}$$

where $\varepsilon_\alpha = \Delta\varepsilon_v/\Delta\log t$, $\varepsilon_\alpha/\varepsilon_c = 0.03$, and $\varepsilon_c = \Delta\varepsilon_v/\Delta\log\sigma'_v$ is determined from:

$$\varepsilon_c = \frac{1.4}{\bar{N}^{1.4}} \tag{5}$$

Average values of DSPT N at each depth from the six profiles were used for all footings, and no energy correction was considered necessary for the measured N values. The predictions of load for 25 mm EOC settlement and settlement 20 years after construction are shown in Table 1.

METHOD USING PCPT q_c (Method B)

Settlement of a footing on sand consists of EOC s_c and post-construction s_s. The method for computing s_c, using Push Cone Penetration Test (PCPT) q_c values, is patterned after the work of Schmertmann (1970). The settlement equation for an embedded footing on sand is:

$$s_c = \left(q - \sigma'_{vo}\right) \sum_{j=1}^{n} \left[Z_I \frac{I'_z}{E_s}\right]_j \tag{6}$$

where E_s is the drained Young's modulus of sand. The value of vertical strain influence factor I'_z for embedded footings is obtained from I_z in Fig. 1 for footings that are placed at the ground surface, using the correction factor in the inset. Values of $[I'_z]_j$ and $[E_s]_j$ are evaluated at middepth of sublayer j. The value of $Z_I = \sum [Z_I]_j$ is computed from:

$$\frac{Z_I}{B} = 2\left(1 + \log \frac{L}{B}\right) \tag{7}$$

Equation 7 is valid for length, L, to breadth, B, ratios (L/B) between 1 and 10.

The value of E_s is obtained from PCPT tip resistance, q_c. The empirical relationship between modulus E_s and cone resistance q_c, based on settlement of foundations and plate load tests, for circular or square loading conditions is:

$$E_s = 3.5\ \bar{q}_c \tag{8}$$

where $\bar{q}_c$ is the weighted mean of the measured PCPT q_c values of sublayers within the Z_I defined by Eq. 7:

$$\bar{q}_c = \sum_{j=1}^{n} \frac{[Z_I]_j}{Z_I} [q_c]_j \tag{9}$$

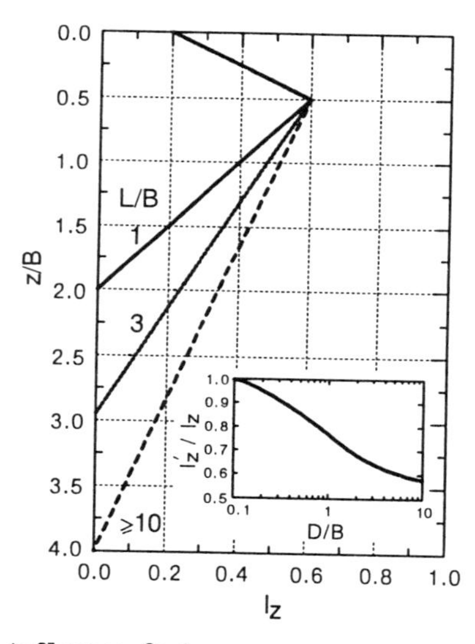

Fig. 1. Strain influence factors.

Post-construction settlement is computed using Eq. 4, whereas, Z_I is determined using Eq. 7, and the value of ε_c is obtained from:

$$\varepsilon_c = \frac{0.1}{\bar{q}_c} \tag{10}$$

where $\bar{q}_c$ is in MPa.

Average values of PCPT q_c at each depth from the five profiles were used for all of the footings. The predictions of load for 25 mm EOC settlement, and settlement 20 years after construction are shown in Table 1.

Table 1. Load and Settlement Predictions

Footing Width, m	1 3 x 3		2 1.5 x 1.5		3 3 x 3		4 2.5 x 2.5		5 1 x 1	
Method	A	B	A	B	A	B	A	B	A	B
P for 25 mm (kN)	3600	3050	1500	1450	3600	3050	2850	2700	950	900
P for 150 mm										
$\Delta s_{1\text{-}30\ min}$										
$s_{20\ year}$, mm	31	34			31	34				

Note that the settlement level specified by the organizing committee does not include settlement resulting from the weight of 1.2 m thick concrete footing.

ACKNOWLEDGEMENT

Complete details of methods A and B for settlement analyses of foundations on granular soils, including overconsolidated sands, surface or embedded foundations, and foundations with length to breadth ratios greater than one, appear in the third edition of Soil Mechanics in Engineering Practice, which is in preparation.

REFERENCES

Burland, J.B., and Burbidge, M.A. (1985). "Settlement of foundations on sand and gravel." Proc. Instn. Civil Eng. Ground Eng. Group Part 1, Dec., pp. 1325-1381.

Gibbens, R., and Briaud, J.-L. (1993). "Data and prediction request for the spread footing prediction event." Report prepared for Settlement '94, Texas A&M University, 71 p.

Schmertmann, J.H. (1970). "Static cone to compute static settlement over sand." *Journal of Soil Mechanics Foundation*, ASCE, Vol. 96, No. SM3, pp. 1011-1043.

SETTLEMENT '94

Franco Ariemma, Senior Soils Engineer
Avtar S. Vasu, Junior Soils Engineer
Fred Agostino, Assistant Soils Engineer
Judith S. Doyle, Assistant Soils Engineer

This report presents the prediction of settlement for five spread footing foundations located at Texas A&M University, Riverside Campus. A conference, sponsored by the FHWA for the ASCE, is scheduled in June of 1994 to discuss the prediction and the actual settlements measured after each footing is load tested.

The general layout of the foundations is given together with laboratory test results from Triaxial Tests, Resonant Column Tests and Index Property Tests. Various insitu tests were performed and the results were mailed to the predictors. These tests include: Cone Penetration Test (CPT), Standard Penetration Test (SPT) with energy, Pressuremeter Test (PMT), Dilatometer Test (DMT) with axial thrust, and others. Foundation embedment depth varies between 0.711 m and 0.889 m. The groundwater table is assumed to be at 4.9 m below the ground surface and the average natural water content is 5%, thus the soil is not fully saturated at foundation depth.

All settlement calculations were done by distributing the pressure to an effective depth of twice the width of the footing. This assumption is based on stress distribution from Boussinesq theory of elasticity and strain influence factor distribution from the cone penetrometer theory. The stress applied at foundation depth reduces to only about 10% at a depth of twice the footing width and the strain influence factor reduces to about 2% of the maximum strain influence factor based on the cone.

The order in which the foundations will be load tested was also considered because of possible load interaction between the footings. It was determined that some load interaction exists between Footings 1, 3 and 4, but the influence was computed to be very small and therefore was neglected.

No information on pore water pressure dissipation against time was available from any of the tests, thus the problem of predicting primary and secondary settlements against time was rendered difficult. The time rate of dissipation of pore pressure and quantifying the primary and secondary settlements were our prime tasks.

From the given grain-size distribution curves at various depths, an average grain-size curve was selected for the soil within the zone of influence. A sample of sand with that grain-size distribution was recreated in the laboratory and a set

Authors Address: N.Y.S.D.O.T.
Soil Mechanics Bureau
1220 Washington Avenue, State Campus
Albany, NY 12232

of consolidation tests were performed to obtain a rate of dissipation of pore pressure.

Rate of consolidation was also estimated from the given stress-strain relationships from triaxial test results by finding the coefficient of volume compressibility (m_v). Least drainage path was estimated by considering three-dimensional dissipation of pore water pressure. An average permeability was determined for the soil.

The instantaneous settlement was estimated under Footing 1 from CPT, PMT, SPT with energy and DMT with axial thrust, since all these test results were available either under or very near to this footing location. Primary settlement due to one-dimensional compression was also estimated based on the laboratory test results. All the analyses performed did show an excellent agreement in quantifying the primary settlement.

Total settlement under Footing 1 was estimated for year 2014 from CPT and PMT. From the PMT primary and secondary settlement was computed for Footing 1, assuming the secondary consolidation is completed in the 20 year time period. From the CPT total settlement in year 2014 was also computed and compared well with the pressuremeter results. Creep factors for 1 minute and 30 minutes were then backfigured from the CPT based on the creep factor for the year 2014. The percentage of creep settlement taking place at 1 minute and 30 minutes was then estimated from CPT and was found to be 4% and 24%, respectively. Again the settlements predicted from all these tests were used to quantify primary and secondary settlement components at the beginning and at the end of the 20 year period.

The settlements for the rest of the footings were predicted from CPT, PMT and SPT results. The test data from the CPT were used for all predictions because a CPT was performed either right under the footing or very near to it and thus could pick up better the local layering effects and any possibility of a compressible layer. The percentages of different settlement components for other footings were taken to be the same as under Footing 1 because a complete picture was available from different test results at this site. Bearing capacity of soil under each footing was then calculated to check for the possibility of failure of soil before reaching 150 mm of settlement. Engineering judgement was used to finalize the settlements for 1 minute, 30 minutes and 20 years. The results are shown in Fig. 1 and Table 1 as appended herewith.

Acknowledgements -

The authors would like to express their sincere thanks to the Soil Mechanics Laboratory and the General Soils Laboratory at NYSDOT, Soil Mechanics Bureau, for extending their support of providing all the testing facilities. The authors would also like to express their gratitude to the colleagues at the Soil Mechanics Bureau for all their help.

References

Riaund, Jean-Louis and Miran, Jerome, 1992, "The Cone Penetrometer Test," Publication No. FHWA-SA-91-043

"Soil Mechanics, Design Manual 7.1," Publication No. NAVFAC DM-7.1, Department of the Navy, Naval Facilities Engineering Command, May 1982

Baguelin, F., Jezequel, J. F., Shields, D.H., "The Pressuremeter and Foundation Engineer," Trans Tech Publications, 1978

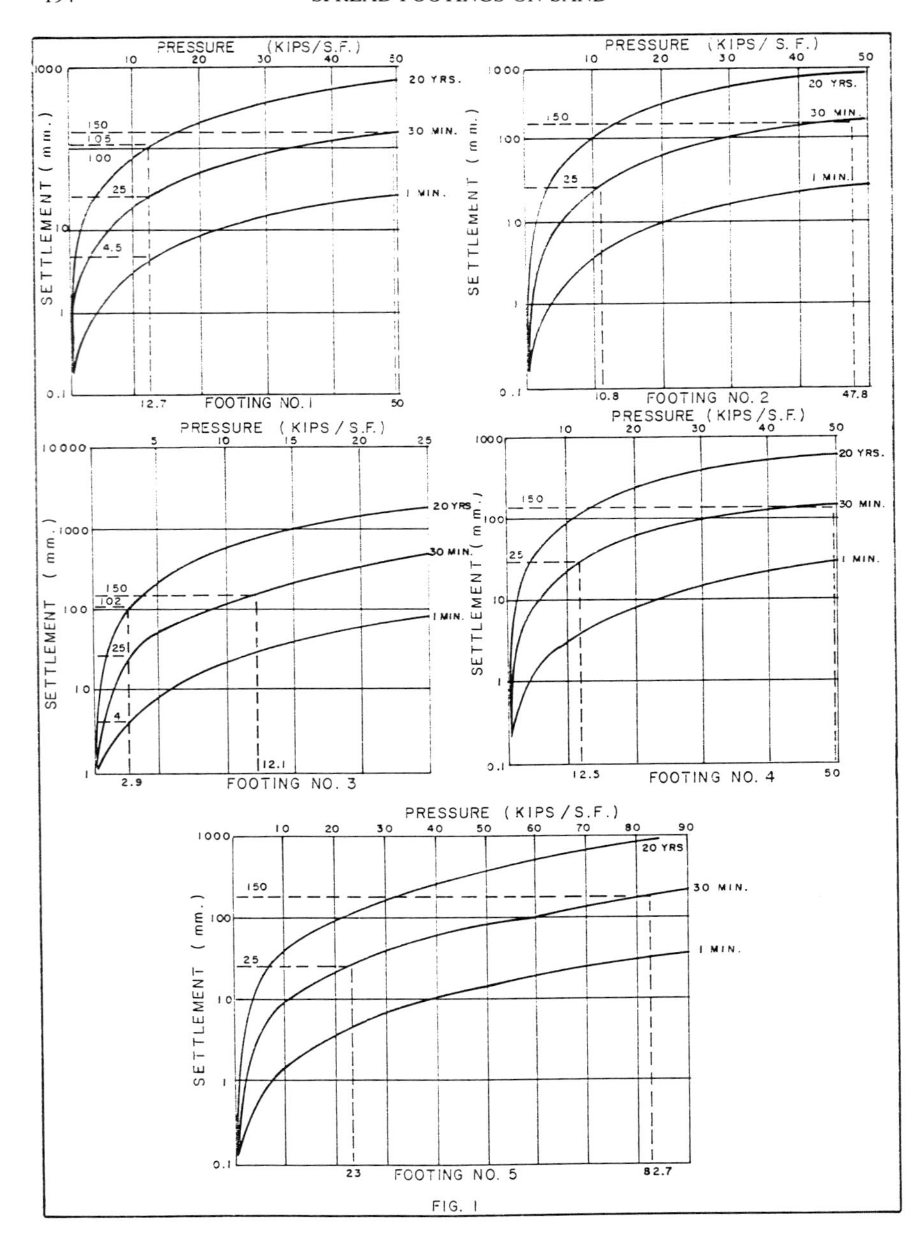
PRESSURE (KIPS/S.F.)
SETTLEMENT (mm.)
20 YRS.
30 MIN.
1 MIN.
150
105
100
25
4.5
12.7
FOOTING NO. 1
10.8
47.8
FOOTING NO. 2
102
4
2.9
12.1
FOOTING NO. 3
12.5
FOOTING NO. 4
23
82.7
FOOTING NO. 5

FIG. 1

	Footing 1 3 m x 3 m	Footing 2 1.5 m x 1.5 m	Footing 3 3 m x 3 m	Footing 4 2.5 m x 2.5 m	Footing 5 1 m x 1 m
Load for 25 mm of settlement Q25 on the 30 minute load settlement curve (Kn)	5480 KN	1165 KN	1250 KN	3750 KN	1100 KN
Load for 150 mm of settlement Q150 on the 30 minute load settlement curve (Kn)	21575 KN	5155 KN	5220 KN	14980 KN	3965 KN
Creep settlement between 1 minute and 30 minute for Q25, Δs (mm)	20.5 mm		21.0 mm		
Settlement in the year 2014 under Q25 (mm)	105.0 mm		102.0 mm		

TABLE 1

Settlement Predictions
Footing Load Tests on Sand

By Kenneth E. Tand[1] and Pickett Warden[2]

The authors have chosen the Burland and Burbidge (1985) method for predicting settlement for normally consolidated sand:

$$S_i = q' B^{.7} I_c = \frac{1.7 q' B^{.7}}{\bar{N}^{1.4}}$$

I_c is compressibility index.
q′ is average gross bearing pressure (kN/m²).
B′ is width of footing (m). Apply correction factor for rectangular footing.
S_i is immediate settlement (mm).
$\bar{N}$ is the average standard penetration number for the depth of influence (Z_1).

The subsoil stratigraphy estimated from the published test data is summarized below:

Depth m	Soil type	N
0 - 4	Medium dense sand	17
4 - 7.5	Medium dense sand	19
7.5 - 11	Hard clay	52

The relative density of the sand deposit is estimated to be in the order of 55 percent. The author's knowledge of the area geology suggests that the sands have been densified by alternating cycles of water saturation. However, it is doubtful that the sands are significantly

[1] Principal Engineer, Kenneth E. Tand & Associates, 1408 E. North Belt, Suite 150, Houston, Texas 77032

[2] Project Engineer, Kenneth E. Tand & Associates, 1408 E. North Belt, Suite 150, Houston, Texas 77032

overconsolidated because the relative density is judged to be relatively low. Thus, the sand deposit was assumed to be normally consolidated for the purpose of the settlement predictions.

Time (min)	S_t/S_1 min
1	1
2	1.027
4	1.087
8	1.177
15	1.224
30	1.289

The settlement ratio was used to estimate the creep settlement from time t = 1 minute to 30 minutes. However, it should be noted that the test footings Tand et. al. references were underlain by a .5 meter layer of stiff sandy clay. Some of the measured settlement with time could be due to volume changes in the clay. Layers of sandy clay are also interbedded in the sand at the Texas A&M test site, but occurred at depths below 5 meters. Some settlement will occur in these layers during load testing of the larger footings.

Burland and Burbidge (1985) and Schmertmann (1972) predict that long-term settlement of foundations on sand will be approximately 1.5 times the initial settlement. However, Gifford et. al. (1987) found that long-term settlement (creep) of footings on sand was minimal based on the measured performance of several bridge foundations. Monitored settlement of a mat and tank bearing on sand at a petrochemical refinery in the Texas Gulf Coast area indicates that the rate of settlement was not measurable after 1 year. For purpose of settlement prediction, the authors assume that settlement after 20 years would be 1.5 times the initial settlement.

The load tests by Tand et. al. indicate that bearing failure occurred at a ratio of settlement to footing width (S/B) of .05. Therefore, the load predicted for S = 150 mm for the footings at the Texas A&M test site assumed that failure conditions had occurred. The ultimate bearing pressure was computed based on Menard's (1962) method as revised by Briaud (1986).

$$q_u = kP_l$$

References

Burland, J. B. and Burbidge, M. C. (1985), "Settlement of foundations on sand and gravel", *Proceedings of the Institution of Civil Engineers*, **78**, Part 1.

Gifford, D. G., Kraemer, S. R., Wheeler, J. R. and McKown, A. F., (1987), "Spread Footings for Highway Bridges", (Report No. FHWA/RD-86/185).

Menard, L. (1965), "Rules for the Calculation of Bearing Capacity and Foundation Settlement Based on Pressuremeter Tests", *6th ICSMFE*, (Vol. 2), 295-299.

Schmertmann, J. H. (1978), "Guidelines for Cone Penetration Test Performance and Design," *Federal Highway Administration*, (Report No. FHWA-TS-78-209).

Tand, K.E., Funegard, E., and Warden, P. (1994), "Footing Load Tests on Sand", Proceeding of Settlement '94.

	Footing 1 3 m x 3 m	Footing 2 1.5 m x 1.5 m	Footing 3 3 m x 3 m	Footing 4 2.5 m x 2.5 m	Footing 5 1 m x 1 m
Load for 25 mm of settlement Q25 on the 30 minute load settlement curve (Kn)	3700	1550	3700	3000	850
Load for 150 mm of settlement Q150 on the 30 minute load settlement curve (Kn)	8740	2170	8830	6020	960
Creep settlement between 1 minute and 30 minute for Q25.Δs (mm)	2.8		2.8		
Settlement in the year 2014 under Q25 (mm)	38		38		

SETTLEMENT PREDICTIONS
FOOTING LOAD TESTS ON SAND

By Erik Funegard[1] and Pickett Warden[2]

The authors have chosen Menard (1965) pressuremeter method for predicting settlement:

$$S = \frac{2}{9E_d} q B_O \left(\lambda_d \frac{B}{B_O} \right)^{\alpha} + \frac{\alpha}{E_c} q \lambda_c B$$

The settlement calculation's were performed using the computer program SHALPMT developed at Texas A&M University by Tucker and Briaud (1986). The following values for the pressuremeter modulus (E_O) and limit pressure (P_l) were interpreted from the published test data:

Depth -m-	Soil Type	E_O kN/m²	P_l kN/m²
0 - 4	Medium dense sand	8,200	800
4 - 7.5	Medium dense sand	13,000	1,175
7.5 - 11	Hard clay	153,000	4,300

An alpha (α) value of .5 was selected for overconsolidated sand with $E_O/P_l \sim 12$. Also, analysis of footing load tests on sand by Tand et. al. (1994) indicate that $\alpha = .33$ for normally consolidated sand underestimates settlement. The following ratios of settlement with time were obtained from analysis of the load tests data by Tand et. al.

---

[1] Geotechnical Consultant, Amoco Corporation, P.O. Box 3011, Naperville Illinois 60566

[2] Project Engineer, Kenneth E. Tand & Associates, 1408 E. North Belt, Suite 150, Houston, Texas 77032

Time (min)	S_t/S_1 min
1	1
2	1.027
4	1.087
8	1.177
15	1.224
30	1.289

The settlement ratio was used to estimate the creep settlement from time t = 1 minute to 30 minutes. However, it should be noted that the test footings Tand et. al. references were underlain by a .5 meter layer of stiff sandy clay. Some of the measured settlement with time could be due to volume changes in the clay. Layers of sandy clay are also interbedded in the sand at the Texas A&M test site, but occurred at depths below 5 meters. Some settlement will occur in these layers during load testing of the larger footings.

Burland and Burbidge (1985) and Schmertmann (1972) predict that long-term settlement of foundations on sand will be approximately 1.5 times the initial settlement. However, Gifford et. al. (1987) found that long-term settlement (creep) of footings on sand was minimal based on the measured performance of several bridge foundations. Monitored settlement of a mat and tank bearing on sand at a petrochemical refinery in the Texas Gulf Coast area indicates that the rate of settlement was not measurable after 1 year. No analytical method exists to predict long-term settlement with a high degree of accuracy. For design, the authors would normally assume that settlement after 20 years would be about 1.3 times the initial settlement for footings on sand.

The load tests by Tand et. al. indicate that bearing failure occurred at a ratio of settlement to footing width (S/B) of .05. Therefore, the load predicted for S = 150 mm for the footings at the Texas A&M test site assumed that failure conditions had occurred. The ultimate bearing pressure was computed based on Menard's (1962) method as revised by Briaud (1986).

$$q_u = kP_l$$

References

Gifford, D. G., Kraemer, S. R., Wheeler, J. R. and McKown, A. F., (1987), "Spread Footings for Highway Bridges", (Report No. FHWA/RD-86/185).

Menard, L. (1965), "Rules for the Calculation of Bearing Capacity and Foundation Settlement Based on Pressuremeter Tests", *6th ICSMFE*, (Vol. 2), 295-299.

Schmertmann, J. H. (1978), "Guidelines for Cone Penetration Test Performance and Design," *Federal Highway Administration*, (Report No. FHWA-TS-78-209).

Tand, K.E., Funegard, E., and Warden, P. (1994), "Footing Load Tests on Sand", Proceeding of Settlement '94.

	Footing 1 3 m x 3 m	Footing 2 1.5 m x 1.5 m	Footing 3 3 m x 3 m	Footing 4 2.5 m x 2.5 m	Footing 5 1 m x 1 m
Load for 25 mm of settlement Q25 on the 30 minute load settlement curve (Kn)	4470	1570	4470	3370	850
Load for 150 mm of settlement Q150 on the 30 minute load settlement curve (Kn)	8740	2170	8830	6020	960
Creep settlement between 1 minute and 30 minute for Q25.Δs (mm)	2.8		2.8		
Settlement in the year 2014 under Q25 (mm)	32		32		

PREDICTION OF VERTICAL LOAD ON SPREAD FOUNDATIONS AT SMALL AND LARGE DEFLECTIONS

Richard J. Deschamps,[1] Associate Member, ASCE, and Scott J. Ludlow,[2] Member, ASCE

INTRODUCTION

This paper outlines the authors' procedures and predictions for the spread footing prediction event held in conjunction with the ASCE Specialty Conference: Settlement '94. Predictions are made of the vertical load required to induce settlements of 25 and 150 mm for five square spread footings ranging in size between 1 and 3 meters. Predictions are also made of the creep settlement between a 1 minute and 30 minute time interval and total accumulated settlement from the time of the initial loading to the year 2014 for a 3 m footing. All predictions have been made well in advance of the measured results being known.

A report of geotechnical data from both in-situ and laboratory tests was made available to all participants (Gibbens and Briaud, 1993). The predictions are based primarily on engineering judgement after considering several empirical methods of analysis (Knowles, 1990; Leonards and Frost, 1989) and results of finite element modeling. The numerical analysis was used primarily for guidance in estimating loads at 150 mm settlements because little experience is available at these deformations.

SETTLEMENT CONSIDERATIONS

Three factors are thought to have an important influence on the resulting load-deformation characteristics at this site.

1) The granular soil has been prestressed. The actual degree of pre-stress is difficult to estimate; however, the geologic formation of the site as a coastal plain indicates significant pre-stress is possible due to wave action, water level fluctuation and desiccation (Briaud, 1993). As a minimum 1.5, meters of soil was removed as part of the grading operations (Gibbens and Briaud, 1993).

2) The bearing soil has an appreciable amount silt content. During the dry summer months when the footing load tests will be performed, the soil is likely to be much stiffer

---

[1]Assistant Professor, School of Civil Engineering, Purdue University., West Lafayette, IN 47907.

[2]Principal, Earth Exploration, Inc., 5958 W. 71st St., Indianapolis, IN 46278, and Ph.D. Candidate, School of Civil Engineering, Purdue University, West Lafayette, IN 47907.

than indicated by the in situ tests taken during the "wetter" spring months.

3) Many empirical settlement procedures were "calibrated" using case histories that relate foundation deflection measurements to "design" loads. A limitation of this approach is that the actual load on the footing is not known, and the design load may be conservative.

PREDICTION PROCEDURES

Conventional settlement analyses were performed to estimate the load for 25 mm deflection using the computer program CSANDSET (Knowles, 1990). This program estimates settlement using fifteen common methods of analysis. As expected, a very broad range in calculated settlements were obtained between the various methods; typically an order of magnitude.

The method proposed by Leonards and Frost (1988) was also considered. This method uses cone penetration and dilatometer test results, along with empirical relationships for K_o and OCR to estimate the magnitude of pre-stress. Young's modulus is estimated from the dilatometer E_d parameter and the recompression modulus is increased by a minimum factor of 3.5. Settlements are calculated using the strain influence diagram proposed by Schmertmann (1970). Two factors contributed to somewhat greater uncertainty for this method at this site: 1) pre-boring was required to advance the dilatometer (Gibbens and Briaud, 1993) causing some stress relaxation, 2) a significant increase in stiffness of the granular soil is possible between the times of in situ testing and foundation loading due to seasonal variation in moisture content.

A finite element analysis was used to provide guidance for estimating loads at 150 mm of deflection because conventional settlement procedures were considered inappropriate due to nonlinearity in the load-deformation curve. A nonlinear elastic, stress dependent, hyperbolic soil model (Duncan and Chang, 1970) was used in the program CRISP (Britto and Gunn, 1990) to perform an axisymetric finite element simulation of the footing load versus deformation. Model parameters were developed from consolidated-isotropically, drained compression (CIDC) test data (Gibbens and Briaud, 1993) and were subsequently modified to account for aging/natural fabric (Schmertmann, 1991). Both the original and modified hyperbolic model parameters are shown in Table 1.

The creep settlements were estimated using the concept of constant C_α/C_c (Mesri and Godlewski, 1977) with a few gross assumptions to obtain C_c and the load increment.

Table 2 provides a summary of the estimated loads for 1.5 and 3 m footings based on the previously discussed procedures. Included in this table are: 1) the average and median estimates of load at 25 mm settlement based on the fifteen settlement procedures available in the CSANDSET program (Knowles, 1990), 2) estimates of load at 25 mm settlement based on the procedure of Leonards and Frost (1988) using the interpreted OCR profile and an assumed OCR profile that results from removal of 3 meters of soil, 3) estimates of loads at 25 and 150 mm of settlement based on the axisymetric finite element analysis using modified parameters.

LOAD PREDICTIONS AND SUMMARY

Typically, when designing shallow foundations, a maximum allowable bearing pressure is recommended that will conservatively limit settlement to some prescribed value (eg. 25 mm). Seldom do actual settlements approach the limiting value. The viewpoint for this prediction event is much more challenging since the goal is to predict the actual applied load that will induce a specific settlement.

Predictions of the load required to induce 25 and 150 mm of settlement in each of the five footings are shown in Table 3. These predictions were made after considering both: the estimates of settlement from the discussed procedures, and the factors listed above which are believed to influence the magnitude of settlement at the site. The predicted loads are much larger than typical recommended design loads when the specified deformations are limiting values.

REFERENCES

Briaud, J.L. (1993). Personal Communication

Britto, A.M. and Gunn, M.J. (1990) "CRISP 90 - User's and Programmer's Guide," Cambridge University.

Duncan, J.M. and Chang, C.Y. (1970). "Nonlinear Analysis of Stress and Strain in Soils," Journal of Soil Mechanics and Foundations Division, ASCE, Vol. 96, No. SM5.

Gibbens, R. and Briaud, J.L. (1993). "Data and Prediction Request for the Spread Footing Prediction", Texas A&M Report.

Knowles, V.R. (1990). "CSANDSET - Settlement of Shallow Footings on Sand," US Army Engineers Waterways Experiment Station, (ITL-91-1)

Leonards, G.A. and Frost, J.D. (1988). "Settlement of Shallow Foundations on Granular Soils," Journal of Geotechnical Engineering, ASCE, Vol. 114, No. 9.

Mesri, G. and Godlewski, P.M. (1977). "Time- and Stress-Compressibility Inter-relationship," Journal of Geotechnical Engineering, ASCE, Vol. 103, GT5.

Schmertmann, J.H. (1991). "The Mechanical Aging of Soils," Journal of Geotechnical Engineering, ASCE, Vol. 117, No. 9.

Schmertmann, J.H. (1970). "Static Cone to Compute Settlement Over Sand," J. Soil Mech. & Found. Div., ASCE, Vol. 96, No. SM3.

Table 1. Hyperbolic Model Parameters

	K	K_b	ϕ	$\Delta\phi$	R_f	n	m
Original	450	430	37	0	0.95	0.30	0.68
Modified	675	645	37	2	0.95	0.30	0.68

Table 2. Summary of Estimated Loads from Various Procedures

Method	Estimated Load (kN)	
	1.5 m x 1.5 m	3.0 m x 3.0 m
Average value from 15 settlement methods (Knowles, 1990) (25 mm)	1030	2590
Median value from 15 settlement methods (Knowles, 1990) (25 mm)	1230	3100
Leonards and Frost procedure with interpreted OCR profile (25 mm)	1700	6250
Leonards and Frost procedure with OCR profile due to 3 m soil removal (25 mm)	1720	4590
Finite element analysis (25 mm)	720	2340
Finite element analysis (150 mm)	2500	8550

Table 3. Footing Load Predictions at 25 and 150 mm Deflections

	Footing 1 3 m sq.	Footing 2 1.5 m sq.	Footing 3 3 m sq.	Footing 4 2 m sq.	Footing 5 1 m sq.
Load for 25 mm of settlement, Q25 on the 30 minute load settlement curve (kN)	5200	1800	5200	2700	900
Load for 150 mm of settlement, Q150 on the 30 minute load settlement curve (kN)	10800	3400	10800	5400	1500
Creep settlement between 1 minute and 30 minute for Q25, Δs (mm)	1.3		1.3		
Settlement in the year 2014 under Q25 (mm)	30.0		30.0		

Prediction of Settlement for Five Footings

Ameir Altaee* and Bengt H. Fellenius, M.ASCE**

Introduction

The subject prediction exercise is very timely, coinciding with recent demands on the geotechnical engineer to provide more exacting information on the settlements affecting a design. Current practice normally considers settlement of footings in sand to be negligibly small, "one inch or smaller", provided, of course that the imposed (the allowable) stresses are kept reasonably conservative. However, this vague and highly qualitative approach is not always safe for the foundation, nor, therefore, for the designer responsible for the performance of the design. No doubt the prediction effort will include a progression of methods, such as rules of thumb, empirical relations, influence factors, and similar approaches of mostly local value, as well as the more rigorous methods of recent development. The authors' prediction is based on a finite element code that incorporates a bounding surface plasticity model for sand.

Soil Constitutive Model

The stress-strain response of the sand is determined by means of a bounding-surface plasticity model, developed by Bardet (1986) and modified by Altaee (1991). The model was implemented by Altaee (1991) into a finite element program, the AGAC93, which has been used successfully in the analysis of several boundary value problems, including full-scale piles driven in sand and subjected to repeated axial loading tests (Altaee et al., 1992a; 1992b; 1993) and Altaee and Fellenius (1993).

The soil constitutive model builds on the principles of steady state soil mechanics. Important features of sand behavior can be modeled, such as stress-hardening, strain-softening, and accumulation of irrecoverable strains during cyclic loading.

The model requires nine parameters for modeling soil behavior in generalized three-dimensional conditions as listed in Table 1. The first four parameters are the most important of the nine indicated. Details of the model formulation, the procedures for determining the parameters, and the finite element program are given by Altaee (1991).

Calibration of the Soil Model

The information provided to the participants in the prediction effort consisted of laboratory and field data. In a preliminary estimation of the nine soil parameters, the authors made use of the basic soil descriptions, such as grain size and in-situ void ratio ('natural void ratio'), and the results of the cone penetration tests. Thereafter, three triaxial tests on reconstructed samples were simulated by means of the AGAC93 program with the estimated parameters as input. The results were reviewed and the differences

---

* Anna Geodynamics Inc., 5350 Canotek Road, Unit 22, Ottawa, Ontario, K1J 9E2

** University of Ottawa, Department of Civil Engineering, 161 Louis Pasteur St., Ottawa, K1N 6N5 and Anna Geodynamics Inc., 5350 Canotek Road, Unit 22, Ottawa, Ontario, K1J 9E2 between experimental and simulated responses were used toward fine-tuning the soil parameters for a renewed program run. After a few trial runs, an acceptable agreement was achieved for both the Deviator Stress versus Axial Strain and Volumetric Strain versus Axial Strain (the information provided to the participants). The results of the final calculations are presented in Fig. 1.

Table 1. Modeling parameters for the bounding-surface plasticity model

Γ	void ratio at 100 kPa mean stress along steady/critical state line (-)	(0.80)
λ	slope of steady/critical state line in the e-ln(p) plane (-)	(0.04)
κ	unloading reloading modulus in the e-ln(p) plane (-)	(0.005)
ϕ_c	ultimate friction angle in triaxial compression (°)	(36.5)
ϕ_e	ultimate friction angle in triaxial extension (°)	(36.5)
ϕ_p	peak friction angle at largest upsilon value in triaxial compresson (°)	(38.0)
ν	Poisson's ratio (-)	(0.35)
ρ	bounding-surface aspect ratio (-)	(2.0)
h_o	hardening parameter	(1.0)

The triaxial test data provided to the participants in the prediction effort do not include any information from the unloading of the tests. It is unfortunate that neither is any information included from results from isotropic compression tests or one-dimensional consolidation tests. Therefore, the laboratory test results provide a calibration of the soil parameters that is less complete than the authors would have desired.

The next step in the calibration process was to apply the soil parameters to a calculation of the field pressuremeter data. To match the unloading behavior, some adjustment was made in the unloading/reloading modulus, κ, but no other changes were found necessary for establishing an agreement between the five simulated and measured pressuremeter tests. This suggests that the sand at the site is isotropic. Fig. 2 shows a comparison between the experimental and simulated pressuremeter curves from three depths: 1.2 m, 3.3 m, and 5.1 m in terms of the Radial Stress versus the Radial Strain measured by the pressuremeter. (The results of the simulation for the intermediate depths are omitted due to lack of space). The zero-value of each pressuremeter test was estimated from the test curves and the simulations have been plotted from this common origin.

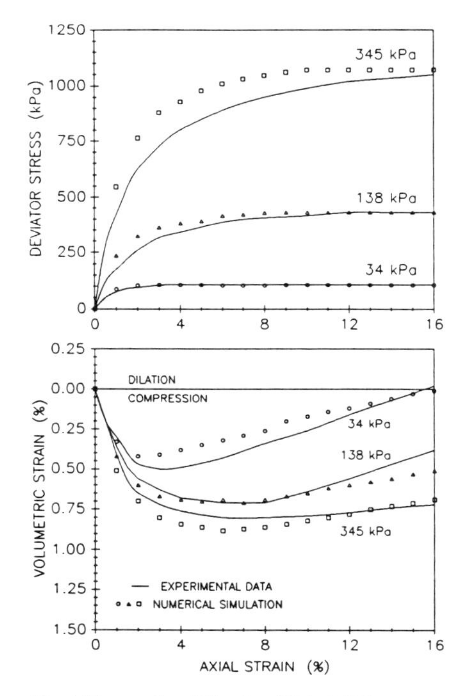

Fig. 1 Measured and Simulated Triaxial Test

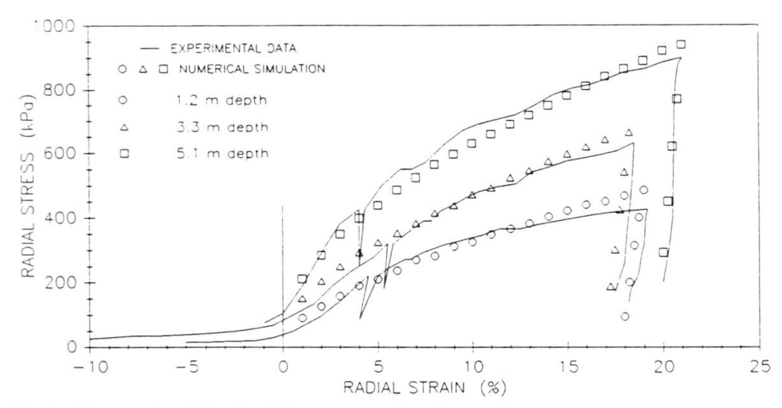

Fig. 2 **Measured and Simulated Pressuremeter Results**

The soil parameters determined from the triaxial test results and the pressuremeter data are listed to the far right in Table 1. To theses values can be added that the pressuremeter data suggest that the sand is overconsolidated and that the K_0 value is 0.7. The in-situ void ratio of 0.75 given for Depth 3.0 m was taken as representative for the site. The compact density indices ('relative density') were not considered.

Prediction of Footing Load-Movement

All five test footings were analyzed by means of the AGAC93 program. The information kit does not indicate the magnitude of load increments used in the static tests. While the prediction kit indicates that an unloading reloading sequence is planned, no information is given with regard to from what load unloading is made. Furthermore, no information is provided on whether the reaction load will be arranged by a loaded platform, by reaction piles, or by other means. Of course, the manner of placing of several hundred tons of reaction load on the ground at the site will have an influence on the load-movement behavior of the footings. As will, for that matter, the sequence of the loading, the existence of the nearby embankment, and the fact that the site has been excavated. These influences are not thought to be major, however. Although the AGAC93 program can include all of these factors, as not all are known, all were omitted from the computations, and the calculations were made rather simply by determining the load that resulted in 5 mm increments of movement, rather than by simulating increments of load.

The authors verified in a separate computer run that one unloading-reloading sequence, if made at low stress level, would result in an about 3 mm increase of movement. This effect would gradually be eliminated at continued loading. If one unloading-reloading sequence would be made at large stress level, it would result in an about 7 mm increase of movement.

Furthermore, it is clear that the true zero (origin) of the test would not be established: Readings of movement would only start after the footings have been cast and the heave due to the excavation of the sand for the placement of the footing and the settlement due to the self-weight of the footing (about 30 KPa) had occurred. Calculations by the AGAC93 calculation suggest that the heave will be about 1 mm and the initial settlement about 2 mm to 3 mm; a smaller value for the small footing and larger for the large footing. However, the authors' prediction data have been related to the same zero situation as used for the tests.

The calculated load-movement curves for the five tests are presented in Figs. 3a and 3b. Table 2 presents the requested prediction information: the footing loads that generate movements of 25 mm and 150 mm. The calculations assume ideal loading (concentric and vertical). Naturally, should a footing tend to tilt during the test, the movement will be affected. The creep calculation has been estimated from the pressuremeter data.

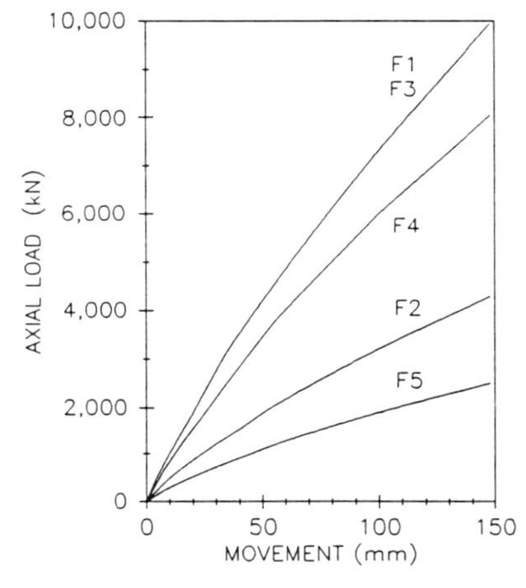

Fig. 3a Computed Load-Movement Curves

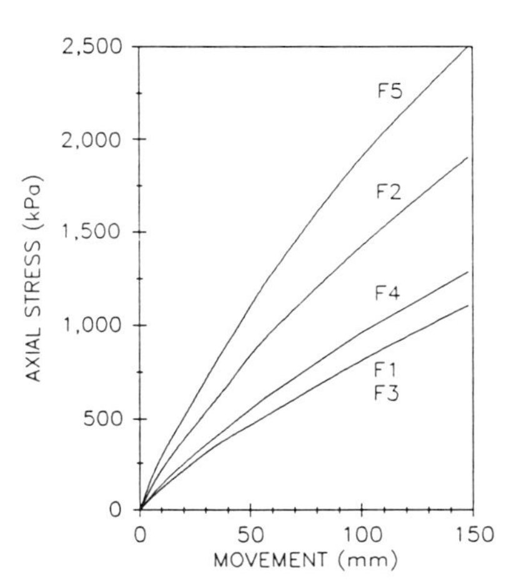

Fig. 3b Computed Stress-Movement Curves

Table 3. Prediction Summary

	F1	**F2**	**F3**	**F4**	**F5**
	3.0 × 3.0	1.5 × 1.5	3.0 × 3.0	2.5 × 2.5	1.0 × 1.0
Q-25 mm (KN)	2,300	1,100	2,300	1,900	600
Q-150 mm (KN)	10,000	4,300	10,000	8,000	2,500
Creep 1-30 (mm)	3		3		

References

Altaee, A., 1991. Finite element implementation, validation, and deep foundation application of a bounding surface plasticity model. Ph.D. thesis, University of Ottawa, Ottawa, Ontario.

Altaee, A. and Fellenius B. H., 1994. Physical modeling in sand. Canadian Geotechnical Journal, Vol. 31.

Altaee, A., Evgin, E., and Fellenius, B. H., 1992a. Axial load transfer for piles in sand, II: Numerical analysis. Canadian Geotechnical Journal, Vol. 29, No. 1, pp. 21-30.

Altaee, A., Evgin, E., and Fellenius, B. H., 1992b. Finite element validation of a bounding surface plasticity model. Computers and Structures, Vol. 42, No. 5, pp. 825-832.

Altaee, A., Fellenius, B. H., and Evgin, E., 1993. Axial load transfer for piles in sand and the critical depth. Canadian Geotechnical Journal, Vol. 30, No. 3, pp. 455 - 463.

Altaee, A. and Fellenius, B. H., 1993. Cyclic performance of an earth fill retention Arctic offshore structure. Proceedings of the 4th Canadian Marine Geotechnical Conference, St.John's, Newfoundland.

Bardet, J. P., 1986. Bounding surface plasticity model for sands. ASME Journal of Engineering Mechanics, Vol. 112, No. EM11, pp. 1198-1217.

PREDICTION OF SETTLEMENT OF SPREAD FOOTINGS

LUCIANO DÉCOURT[1], Fellow ASCE

ABSTRACT: Load-settlement relationships were established for five footings. The settlements were computed for 1 min., 30 min. and 20 years loadings. The framework for these computations was the theory of elasticity, with the elastic modulus determined from in-situ tests via the maximum shear modulus G_o.

SOIL TESTS

A great amount of in-situ and laboratory tests were carried out. Among them, the following were considered for assessing G_o: Cross Hole Tests (CHT), Pressuremeter Tests (PMT), Marchetti Dilatometer Tests (DMT), Standard Penetration Tests (SPT) and Cone Penetration Tests with pore pressure measurements (CPT-U).

ELASTIC MODULUS

For the larger footings presenting safety factors against shear failure of no less than ten, it was assumed that the elastic modulus corresponds to shear strains of 10^{-1}% ($E_{0.1\%}$). It was considered appropriate to assess E via G_o.

G_o VALUES

The basic G_o values were those given by the CHT. Since they involve a much larger mass of soil than any other test, a range of values rather than individual ones were considered. (Figures 1, 2 and 3)

For non plastic soils, the shear modulus for a shear strain of 10^{-1}(%), $G_{0.1\%}$ is 0.2 of G_o, Décourt (1991).The corresponding $E_{0.1\%}$ value is therefore approximately equal to 0.5 G_o.

---

[1]Director of Luciano Décourt Engenheiros Consultores Ltda., Av. Brig. Faria Lima, 1867 - 2º - São Paulo, Brazil - CEP 01451-912

Table 1 shows the correlations used for assessing G_o on the basis of the results of the different tests.

Table 1. Correlations used for assessing G_o

Test	G_o	Reference
PMT	$G_o \cong 2xE_R$ (reload modulus)	this paper
DMT	$G_o/E_D = f(P'_o(DMT)$ and σ'_{vo}	Baldi et al(1989)
SPT	$G_o \cong 5.0(N_{60})$ MN/m^2	Décourt et al 1989
CPT	$G_o \cong 11q_c$(tip resistance)	this paper

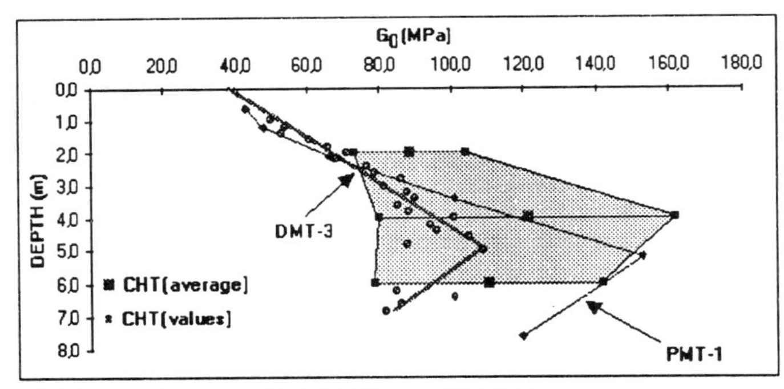

FIGURE 1. G_0 from CHT, PMT-1 and DMT-3

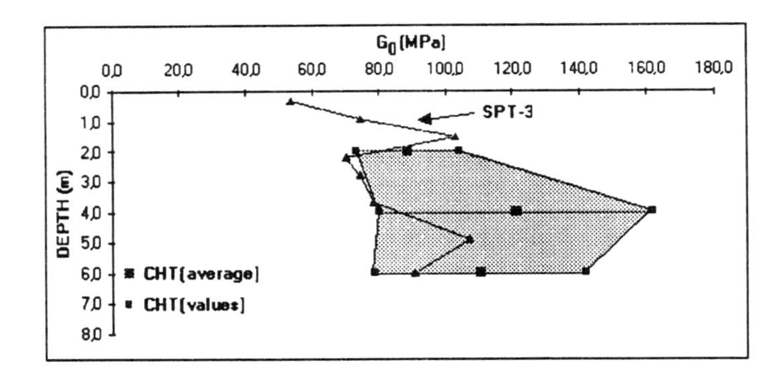

FIGURE 2. G_0 from CHT and SPT-3

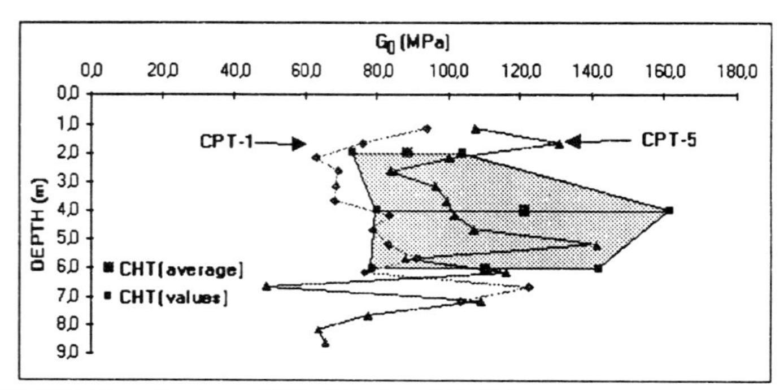

FIGURE 3. G_0 from CHT, CPT-1 and CPT-5

ULTIMATE BEARING CAPACITY

The ultimate bearing capacity is here considered as the load corresponding to a deformation of 10% of the equivalent diameter of the footing. For rigid circular footings with one meter in diameter, Décourt (1991) suggested that the failure stress (σ_r) is equal to N_{60} x 10^2kPa. For any other footing size σ_r is supposed to increase directly with the equivalent diameter (D_{eq}.)

SETTLEMENT OF THE FOOTINGS UNDER THEIR OWN LOADS

The elastic moduli were selected on basis of CPT data corrected by a factor α which is a function of the results of the other tests performed closeby, and a factor β which is a function of the footing size. The overall correction factor $K=\alpha.\beta$ is: 1.0, 1.07, 1.56, 0.98 and 0.675 respectively for footings 1, 2, 3, 4 and 5. The Poisson's coefficient (μ) was considered to be equal to 0.2. The results are presented in Table 2(first lines).

ADDITIONAL SETTLEMENTS CAUSED BY ADJACENT FOOTINGS

The settlements of a given footing depend not only on the loads applied to it but also on the loads applied to the adjacent footings. Since the loading sequence was not clearly established, it is practically impossible to correctly predict the load that will correspond to a given settlement. However, since predictions had to be made some assumptions were considered.

For computational purposes an E value of 50,000 kPa and a μ value of 0.2 were considered.

The settlement influence factors presented in Milovic tables (1992) were used. But, after comparing the values presented in these tables with the experiments by Rocha Filho et. al. (1987) it was decided to halve these influence factors for values of r/R higher than 3(r being the distance between the centers of the footings and R the radius of the footing whose influence is being considered) The following scenario was considered:

Footing 5 is loaded to failure and its influence on the settlement of the others is considered negligible.

Footing 4 is loaded to 3,329.18 kPa, 40% of its failure load estimated to be Q_r = 5,822.96 kPa. This loading produces settlement of 7.6mm on footings 1 and 3 and of 5.24mm on footing 2.

Footing 2 is loaded to 829.82 kPa, 40% of the ultimate load of 2,074.55 kPa. This load induces negligible additional settlements in the other footings.

Footing 1 is loaded to 3,374.96 kPa, 40% of the ultimate load of 8,439.41 kPa. This loading produces an additional settlement of 11mm on footing 3 and of 16mm on footing 2. The results are presented in table 2 (second lines).

PREDICTED LOADS AND SETTLEMENTS

The predicted loads and settlements are presented in Table 2.

Table 2. Predicted loads and settlements

	F. 1	F. 2	F. 3	F. 4	F.5
Q_{25} (kN.)	4,290 2,986	1,295 1,024	4,658 1,192	2,740	779.15
Q_{150} (kN.)	25,740 17,915	4,490 4,218	27,945	16,440	2.360
ΔS_1 (mm)	2.0		2.0		
S_{2014} (mm)	44.25		44.25	$S_2=S_1(1+0.2 \lg t_2/t_1)$	

REFERENCES

Amar, S. , Baguelin, F. , Canepa, Y. and Frank, R. (1994) "Experimental study of the settlement of shallow foundation". To be published in the Proc. of the ASCE Conference, "Settlement 94".

Décourt, L., Belincanta, A., Quaresma Filho, A.R. (1989) "Brazilian experience on SPT" Suppl. contr. by the Braz. Soc. for Soil Mechanics, XII ICSMFE, Rio de Janeiro.

Décourt, L. (1991) "Special problems on foundations". Gen. Report. Proc. IX PAMCSMFE, Vol. 4, Viña del Mar.

Milovic, D. (1992) "Stress and displacements for shallow foundations, Elsevier". Amsterdam.

Rocha Filho, P., Romanel, C. and Brown, P.T. (1987) "Interpretation of surface settlements outside loaded area". Proc. VIII PACSMFE, Vol. 2, pp. 347-356, Cartagena

CPT-BASED PREDICTION OF FOOTING RESPONSE

Paul W. Mayne[1], M. ASCE

The results of cone penetration tests (CPT) have been used to predict the nonlinear load-displacement behavior of spread footing foundations resting on a 10-m thick sand deposit. Initial stiffnesses are evaluated using elastic theory with low-strain moduli estimated from G_{max}-q_c correlations. Ultimate bearing stresses are determined from the Vesic (1975) solution for N_γ factors with operational ϕ' evaluated from the CPT data. A modified hyperbolic relationship is adopted between these extremes.

All five CPT soundings were averaged to give a representative profile of cone tip resistance (q_c) with depth. This q_c(ave) was used to generate interpretative profiles of soil properties of the sand layer for foundation analysis including: unit weight (γ), low-amplitude shear modulus (G_{max}), effective stress friction angle (ϕ'), overconsolidation ratio (OCR), and lateral stress coefficient (K_o). Most of the correlations utilized are based on a compiled CPT database derived from 24 separate series of calibration chamber tests on clean, uncemented, unaged sands that were corrected for boundary effects (Mayne and Kulhawy, 1991).

Initially, a value of γ was estimated from the normalized cone tip resistance, $q_{c1} = (q_c/p_a)/(\sigma_{vo}'/p_a)^{0.5}$ where p_a = a reference stress equal to atmospheric pressure ($\approx$ 100 kPa $\approx$ 1 tsf). The procedure is iterative since an assumed value of γ is required to generate the effective stress overburden profile (σ_{vo}'). The values of dry and wet density (γ/γ_w) depend upon void ratio (e_o) and stress history of the sand and may be estimated from the following empirical equations:

NC Sand: $$e_o = 1.158 - 0.230 \log(q_{c1}) \quad [1]$$

OC Sand: $$e_o = 1.232 - 0.245 \log(q_{c1}) \quad [2]$$

An analysis of the CPT data using a method proposed by Mayne (1991) suggests that the sand is overconsolidated with average OCR $\approx$ 6. The normalized q_{c1} averages about 103, from

1. Associate Professor, School of Civil & Environmental Engineering, Georgia Institute of Technology, Atlanta, GA 30332-0355.

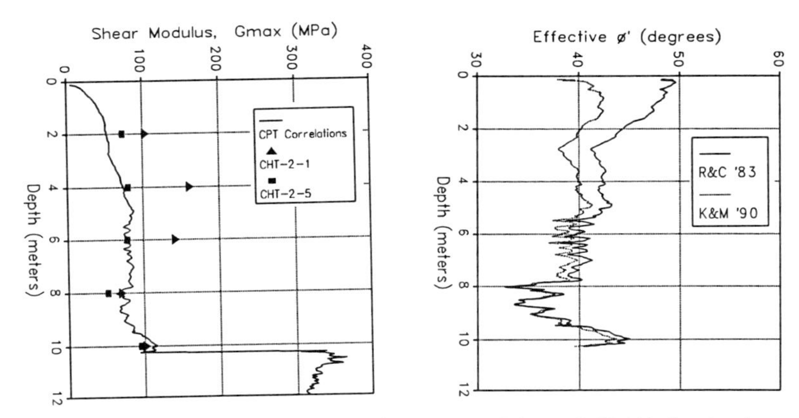

Fig 1. CPT Estimates for (a) Low-Strain Shear Modulus and (b) Friction Angle.

which Eq[2] indicates an in-situ e_o = 0.74. Assuming a specific gravity (G_s = 2.7) gives values of dry and wet unit weights of 15.2 kN/m^3 of 19.3 kN/m^3, respectively, which compare reasonably with the in-situ density measurements reported at the site.

The in-situ stiffness of the sand was evaluated in terms of low-amplitude shear modulus (G_{max}). The empirical expressions utilized include:

Sands: $G_{max} = 1634\ (q_c)^{0.250}(\sigma_{vo}')^{0.375}$ (Rix & Stokoe, 1991) [3]

Clays: $G_{max} = 406\ (q_c)^{1.335}(e_o)^{-1.130}$ (Mayne & Rix, 1993) [4]

where stresses and moduli are in kPa. The derived values of G_{max} are shown in Figure 1a and compare well with the measured results from the crosshole test CHT-2-5 within the sand layer at all depths. The agreement with CHT-2-1 is not very good at shallow depths however. The value of G_{max} in the sand layer averages 70 MPa, and assuming a Poisson's ratio of ν = 0.15, the low-amplitude elastic modulus is taken as E_{max} = 161 MPa.

The effective stress friction angle (ϕ') of the sand layer was evaluated using the following two methods:

$\phi' = \arctan(0.1 + 0.38 \log(q_c/\sigma_{vo}'))$ (Robertson & Campanella, 1983) [5]

$\phi' = 17.6° + 11.0 \log(q_{c1})$ (Kulhawy & Mayne, 1990) [6]

Profiles of the interpreted ϕ' are shown in Figure 1b. Average values of ϕ' = 41.8° and 39.3° were determined, respectively, and the latter value adopted henceforth for analysis.

The ultimate bearing capacity has been determined using the approximate form for shallow foundations given by (Vesic, 1973, 1975):

Sands: $q_{ult} = B\gamma N_\gamma/2$ [7]

where B = footing width and $N_\gamma = 1.2[\exp(\pi\tan\phi')\tan^2(45° + \phi'/2) + 1]\tan\phi'$ for a square footing. The adopted $\phi' = 39.3°$ gives $N_\gamma = 58$. Note that N_γ is highly sensitive to value of ϕ'. The dry γ was used since the groundwater lies approximately 5 m deep. Calculated ultimate bearing stresses for B = 1, 1.5, 2.5, and 3 m are 451, 677, 1128, and 1353 kPa.

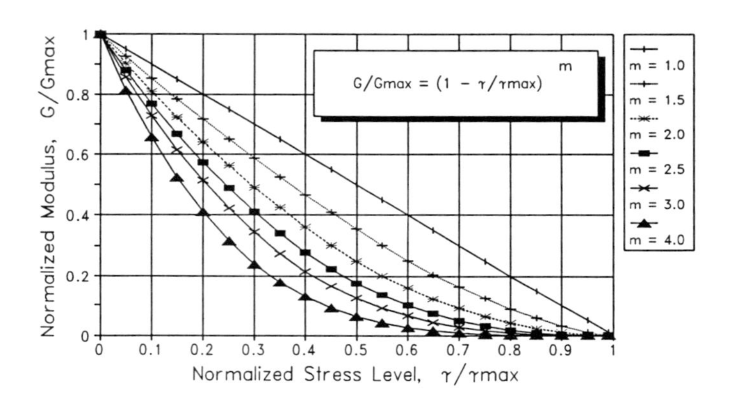

Figure 2. Modulus Degradation Curves from Modified Hyperbola Approach.

Settlements have been calculated using influence factors derived from elastic theory solutions (Poulos and Davis, 1974). The basic format is: $\delta = qBI/E$, where δ = vertical displacement, E = elastic modulus, and I = strain influence factor. Various soil models were tried and eventually a constant E profile was adopted for the 10-m sand layer. Derived values of I ranged from 0.81 to 0.87 for the five square rigid footings (assuming $\nu = 0.15$).

In order to degrade the low-amplitude modulus to appropriate strain levels, a modified hyperbolic form was employed. Recent laborotory studies using local internal strain measurements on sands (Teachavorasinskun, et al. 1991; LoPresti, et al. 1993) have shown that a simple hyperbola overpredicts the monotonic σ-ϵ behavior. A simple adjustment to the normalized modulus degradation ($G/G_{max} \approx E/E_{max}$) with stress level ($q/q_{ult}$) is shown in Figure 2. A review of the available data suggests that a value of m = 2 is appropriate for monotonic loading of OC sands. Therefore, the final predictive equation for foundation settlement is given as:

$$\delta = qBI/[E_{max}(1 - q/q_{ult})^2] \quad [8]$$

where q = applied stress level. Using [9], the predicted nonlinear stress-displacement curves are given on Figure 3. No time-dependency of loading or strain-rate effects have been included in the aforementioned approach.

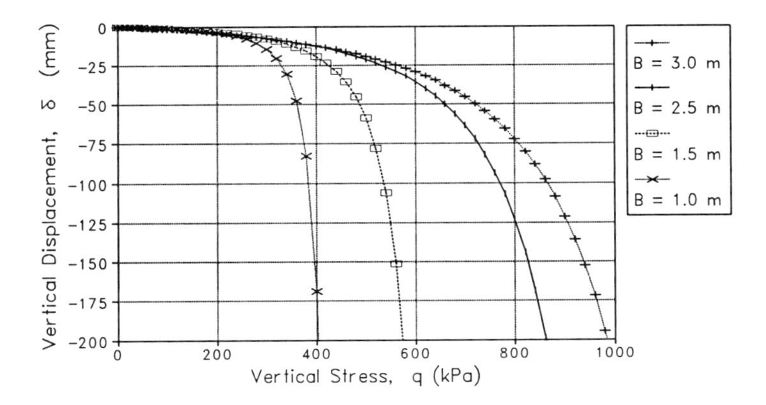

Figure 3. Predicted Vertical Stress-Displacement Response of Footings

REFERENCES

LoPresti, DCF et al. (1993). Monotonic and cyclic loading behavior of two sands at small strains. ASTM *Geotechnical Testing Journal* 16 (4), 409-424.

Kulhawy, FH and Mayne, PW (1990). Manual on estimating soil properties for foundation design. *Report EL-6800.* Electric Power Research Institute, Palo Alto, 306 p.

Mayne, PW (1991). Tentative method for estimating σ_{ho}' from q_c in sands. *Calibration Chamber Testing*, Elsevier, New York, 249-256.

Mayne, PW and Kulhawy, FH (1991). Calibration chamber database and boundary effects correction for CPT. *Calibration Chamber Testing*, Elsevier, New York, 257-264.

Mayne, PW and Rix, GJ (1993). G_{max}-q_c relationships for clays. ASTM *Geotechnical Testing Journal* 16 (1), 54-60.

Poulos, HG and Davis, EH (1974). *Elastic Solutions for Soil and Rock Mechanics*, Wiley & Sons, New York, 411 p.

Rix, GJ and Stokoe, KH (1991). Correlation of initial tangent modulus and cone penetration resistance. *Calibration Chamber Testing* (ISOCCT), Elsevier, New York, 351-362.

Robertson, PK and Campanella, RG (1983). Interpretation of cone penetration tests. *Canadian Geotechnical Journal* 20 (4), 718-745.

Teachavorasinskun, S, Shibuya, S and Tatsuoka, T (1991). Stiffness of sands in monotonic and cyclic torsional simple shear. *Geotechnical Engineering Congress* (GSP 27), Vol. II, ASCE, New York, 863-878.

Vesić, AS (1973). Analysis of ultimate loads of shallow foundations. *Journal of the Soil Mechanics and Foundations Division* (ASCE) 99 (SM1), 45-73.

Vesić, AS (1975). Bearing capacity of shallow foundations *Foundation Engineering Handbook*. Ed. Winterkorn & Fang, Van Nostrand Reinhold Co., New York, 121-147.

Note: B = 2.5m

	Footing 1 3 m x 3 m	Footing 2 1.5 m x 1.5 m	Footing 3 3 m x 3 m	Footing 4 2 m x 2 m	Footing 5 1 m x 1 m
Load for 25 mm of settlement Q25 on the 30 minute load settlement curve (Kn)	5200	950	5200	3350	330
Load for 150 mm of settlement Q150 on the 30 minute load settlement curve (Kn)	8450	1260	8450	5150	395
Creep settlement between 1 minute and 30 minute for Q25, Δs (mm)	NA		NA		
Settlement in the year 2014 under Q25 (mm)	NA		NA		

TABLE 17
Prediction Summary Format

SPREAD FOOTING PREDICTION EVENT

By Ching L. Kuo[1] and Sayed M. Sayed[2]

INTRODUCTION

The behavior of footings on sand at small and large deflections is examined at one of the two National Geotechnical Experimentation sites located on the Texas A & M University Riverside Campus. This prediction event is sponsored by FHWA. The goal of this effort is to access the ability of geotechnical engineers to predict the settlement of shallow foundations on sand.

The data of in-situ tests performed at the site were provided by Texas A & M University. These tests included CPT, PMT, SPT with Energy, DMT with axial force, Step Blade, Borehole Shear and Cross-Hole Wave tests.

The paper presents the predicted behavior based on the analyses performed using the data from CPT, PMT, SPT and DMT.

SOIL AND FOUNDATIONS DATA

Soil

The soil at the National Geotechnical Experimentation Site selected for this project (Gibbens and Briaud 1993) is predominantly sand from 0 to 11 m. Below the sand layers is a clay layer which exists until a depth of at least 33 m. The sand is a medium dense silty fine sand and the clay is a very hard plastic clay. The water table was observed in an open well to be at a depth of 4.9 m. The weather at the site (summer 1993) has been very dry with no rain.

Foundations

Five spread footings were constructed at the site. The footings include two footings with nominal dimensions of 3 m by 3 m, one 2.5 x 2.5 m footing , one 1.5 x 1.5 m and one 1 x 1 m. footing. The footings were founded at a depth of 0.7 to 0.9 m. The "As - built" dimensions vary slightly from those referenced here. All footings were load tested in the Fall of 1993.

PREDICTION METHODS

Cone Penetration Test (CPT)

The method presented by Schmertmann (1970, 1978) and revised by Robertson and Campanella (1988) was used in conjunction with the CPT. The equation used in the analysis is expressed

---

[1] District Geotechnical Engineer, Florida Dept. of Transportation, Bartow, FL 33030
[2] Principal, Geotech Consultants International, Inc. (GCI), Winter Park, FL 32789

as follows:

$$s = C_1 C_2 \Delta_p \Sigma(I_z \Delta z)/(C_3 \alpha q_c) \quad (1)$$

Where Δ_p = net footing pressure, I_z = strain influence factor, Δ_z = soil layer thickness, q_c = cone bearing, α=empirical factor (6 to 10 for OC silica sands), and C_3=shape factor (1.25 for square footing). Also, the correction factors C_1 (depth to base of footing correction) and C_2 (long term creep correction) are expressed as:

$$C_1 = 1 - 0.5(\sigma'/\Delta_p) \ ; \ \sigma' = \text{effective vertical stress at footing depth} \quad (2)$$

$$C_2 = 1 + 0.2 \log(10t) \ ; \quad t = \text{time in years} \quad (3)$$

The nearest CPT sounding was used in predicting the load-settlement behavior for each footing. In the analysis, we used a value of 15 for the parameter $x=C_3\alpha$ to reflect the effect of overconsolidation. We also adjusted the settlement to reflect the stress level by assuming the soil stress-strain behavior in the form of a nonlinear hyperbolic. The adjustment of moduli was made by using $(1-P_b R_b/P_f)^2$ where R_f=failure ratio and was assumed 0.95, P_b=applied foundation load and P_f=ultimate load. In computing the long term settlement , the expression for C_2 was modified to reflect overconsolidation. The modification was based on the ratio of reload/load moduli from the various tests and the coefficient 0.2 in the expression for C_2 was divided by 5, i. e., a value of 0.04 instead of 0.2 was used in the computations. The Prediction Results summarized at the end of this paper are mostly based on the CPT analyses.

Pressuremeter Test (PMT)

The predictions using the Pressuremeter Test results were based on the pressuremeter modulus using a semi-empirical approach proposed by Ménard and Roussean (1962). The distortion deformation s_d and the consolidation (volumetric) deformation s_c are expressed as follows:

$$s_d = (2/9E_M) q^* B_o (\lambda_d B/B_o)^\alpha; \qquad E_M = E_d \quad (4)$$

$$s_c = (\alpha/9E_M) q^* \lambda_c B; \qquad E_M = E_c \quad (5)$$

where E_M is the pressuremeter modulus, q^* is the net average bearing stress, B_o = a reference width, usually 60 cm, B = the width or diameter of the footing, α = rheological factor, which depends on the soil type and the ratio E_M/p_l and p_l is the equivalent net limit pressure, λ_d and λ_c = shape factors, which depend on the length to width ratio of the foundation, the pressuremeter moduli E_d and E_c were determined according to the procedure given by the Centre d'Etudes Ménard (1967).

The computations were based on boring PMT-2. For footings 1, 3 and 4, the moduli E_c and E_d are 9,151 kPa and 11,268 kPa, respectively. The corresponding values for footings 2 and 5 are 10,558 kPa and 9,776 kPa, respectively. The rheological factor α was taken 1/3 (sand with E_M/P_l = 6 to 13) and the values of the shape factors λ_d and λ_c are 1.12 and 1.10 respectively for all footings. The loads for 25 mm and 150 mm of settlement (i.e., Q25 and Q150) on the 30 minute load settlement curves based on the PMT were 2 to 3 times those shown in the summary table of the Prediction Results. The creep settlement between 1 minute and 30 minute for Q25 is 10 mm for footings 1 and 3. The settlement in the year 2014 under Q25 in 25 mm.

Standard Penetration Test (SPT)

Boussinesq solution was used in the computations based on the SPT results. In general, the settlement of a soil mass consisting of a number of layers of different deformation characteristics can be computed from the following expression (Vesic 1978):

$$s = pB\Sigma(I_n - I_{n-1})/E_n^* \quad (6)$$

where $E_n^* = E/(1-\nu^2)$ is the plane- strain modulus of deformation of layer n, E is the modulus of deformation, ν is Poisson's ratio, I_n and I_{n-1} are dimensionless stress factors corresponding to the depths z_n and z_{n-1} respectively, and B is the width of the loaded area. It is implicitly assumed, in Eq. (6), that the difference in rigidity of various layers is small enough not to significantly affect the stress distribution in the mass.

The contribution of an individual layer to consolidation settlement can be found, in principle, by deducting the immediate settlement (obtained from Eq. (6) with undrained deformation characteristics) from the total settlement (obtained from the same equation with drained deformation characteristics).

The loads Q25 and Q150 were computed based on drained parameters (E_d, ν_d). The undrained modulus E_u in kPa was computed from an expression given by Bowles (1988) in the form $E_u = 300\ (N + 6) .\ C$ where N is the SPT blow count within the influence zone of the foundation and the constant C was introduced in this paper to reflect stress level and overconsolidation effects of the silty sand (De Beer 1987; Sayed 1987). Poisson's ratio in undrained and drained conditions were taken 0.5 and 0.33, respectively. Based on the Theory of Elasticity, the drained and undrained moduli are related by the expression $E_d=0.9E_u$.

For Q25, C varied from 2 to 5 and the drained plane strain modulus varied from 14,000 to 39,000 kPa. For Q150, the moduli were determined with C=1. Under Q25, the creep settlement for footings 1 and 3 between 1 minute and 30 minute is 7 mm. The settlement in the year 2014 under Q25 is 25 mm.

Dilatometer Test (DMT)

Schmertmann's Ordinary Method (1986) was used in the predictions based on the DMT results. The expression has the form:

$$s = (\Delta\sigma_i/M_i)h_i \quad (7)$$

where s=settlement, $\Delta\sigma_i$= stress change, M_i= constraint modulus and h_i = layer thickness. The results of DMT-3 were used to predict the load-settlement behavior of footing 3. The moduli values were adjusted to reflect the stress level by using the same factor as the CPT method. The DMT results compare favorably with those shown in the summary table.

SUMMARY

The results of CPT, PMT, SPT and DMT were used to predict the behavior of footings on sand at small and large deflections. The prediction summary is given in the following Table.

Summary of Prediction Results

	Footing 1 3m x 3m	Footing 2 1.5m x 1.5m	Footing 3 3m x 3m	Footing 4 2.5m x 2.5m	Footing 5 1m x 1m
Q25 (kN)	3100	650	2650	2000	320
Q150 (kN)	5500	970	5400	3500	420
Creep (mm)	16	N/A	16	N/A	N/A
Settlement in 2014 under Q25 (mm)	27	N/A	27	N/A	N/A

Notes:
- "As-built" dimensions vary slightly from those shown above;
- Q25= Load for 25 mm of settlement on the 30 minute load-settlement curve;
- Q150= load for 150 mm of settlement on the 30 minute load-settlement curve;
- Creep= settlement between 1 minute and 30 minute for Q25;
- N/A= Not Available (not required)

The 1 minute and 30 minute predictions were made using all methods. the CPT method is the only method that accounts for settlement-time dependence in a direct way. In principle, the predictions presented herein are based on the assumption that the 1 minute predictions correspond to immediate settlement and are mostly due to distortion. The 30 minute predictions are associated with dissipation of excess pore water pressure, i. e. consolidation and is viewed for the most part as the long-term settlement. It is believed that 95 to 98 percent of the settlement would occur in the 30 minute period. In other words, the settlement in the year 2014 will not be significantly different from those at the 30 minute period unless major changes occur in the soil or loading conditions, such as groundwater changes, vibrations as examples.

REFERENCES

Bowles, J. E. (1988). Foundation Analysis and Design, 4th ed., McGraw-Hill Book Co., New York, NY.

De Beer, E. E. (1987). "Analysis of Shallow Foundations." in Geotechnical Modeling and Applications. Sayed, S. M., ed., Gulf Publishing Company, Houston, TX.

Centre d'Etudes Ménard (1967). "Interprétation d'un Essai Pressiométrique", Publication D31/67.

Gibbens, Robert and Briaud, J-L. (1993). "Data and Prediction Request For the Spread Footing Prediction Event." Sponsored by FHWA at the Occasion of the ASCE Specialty Conference: Settlement' 94.

Ménard, L. and Rousseau, J. (1962). "L'évaluation des tassements - Tendances nouvelles" - Sols-Soils, Vol. I, No. 1, Juin, 13-29.

Robertson, P. K. and Campanella, R. G. (1988). "Guidelines for Using the CPT, CPTU and Marchetti DMT for Geotechnical Design." U. S. Department of Transportation, Federal Highway Administration, Office of Research and Special Studies, Report No. FHWA-PA-87-023+84-24.

Sayed, S. M. ed. (1987). Geotechnical Modeling and Applications. Gulf Publishing Company, Houston, TX.

Schmertmann, John H. (1970). "Static Cone To Compute Static Settlement Over Sand." Journal of Geotechnical Engineering Division, ASCE, Vol. 96, SM3, 1011-1043.

Schmertmann, John H. (1978). "Guidelines for Cone Penetration Test, Performance and Design." Federal Highway Administration, Report FHWA-TS-78-209, Washington, D. C.

Schmertmann, John H. (1986). "Dilatometer To Compute Foundation Settlement." Proceedings of ASCE specialty Conference "In Situ 86."

Vesic, A. S. (1978). "Class Notes of CE 235 - Foundation Engineering." Duke University, Durham, NC.

PREDICTIONS OF FOOTING TESTS SPONSORED BY THE FHWA USING A STRAIN HARDENING ELASTOPLASTIC CONSTITUTIVE MODEL

I. SHAHROUR[1], M. ZAHER[2]

INTRODUCTION

A strain hardening elastoplastic constitutive model (MODSOL) is used to predict the footing tests sponsored by the FHWA. After a brief description of this model, we give the methodology followed for the determination of constitutive parameters and the results of numerical predictions.

DESCRIPTION OF THE CONSTITUTIVE MODEL (MODSOL)

MODSOL is a strain haradening elastoplastic constitutive model (Shahrour and Chehade 1992). It is characterized by a single yield surface with isotropic hardening and a non-associated flow rule. The elastic part is assumed to be non linear. Young's moduls and Poisson's ratio are given by :

$$E = E_o\,(p/p_a)^n \quad \text{and} \quad \nu = \nu_o$$

where E_o, ν_o and n are constitutive parameters; p and p_a stand for the mean pressure and a reference pressure, respectively.
The yield surface is assumed to be linear in the (p, q) plane :

$$f = q - M\,R\,p = 0$$

The parameter M depends on both the friction angle (ϕ) and Lode angle (θ). Its expression is due to Zienkiewicz et al (1985) :

$$M = \frac{6\,\mathrm{Sin}(\phi)}{3 - \sin(\phi)\,\sin(3\theta)}$$

The hardening function is assumed to depend on the deviatoric plastic strain:

$$R = \frac{\varepsilon_d^p}{a + \varepsilon_d^p} + b(\varepsilon_d^p)^2 \exp(-c\varepsilon_d^p)$$

where a, b and c are constitutive parameters. The exponential term used in the hardening function allows us to describe strain softening. In the case of a non softening material, constitutive parameters b and c are equal to zero.

The non associated flow rule is derived from the characteristic concept (Luong 1978) and the Cam-Clay model (Schofield and Worth 1968). Dilatancy is controlled through the following condition :

---

[1] Professor, Laboratoire de Mécanique de Lille, Ecole Centrale de Lille, Villeneuve d'Ascq, France.

[2] Research fellow, Laboratoire de Mécanique de Lille, Ecole Centrale de Lille.

$$\frac{d\,\varepsilon_v^p}{d\,\varepsilon_d^p} = (Mg - q/p)e^{-\alpha_g \varepsilon_d^p} \qquad Mg = \frac{6\mathrm{Sin}(\phi_c)}{3 - \sin(\phi_c)\sin(3\theta)}$$

where ϕ_c designates the characteristic angle; α_g is an adjusting parameter which permits to ensure stabilization of the volume change for the high value of the deviatoric plastic strain

DETEMINATION OF CONSTITUTIVE PARAMETERS

Table 1 gives the set of parameters used in the prediction of the footing tests. It was fixed as follows : Elastic parameters were determined from resonant column and cross-hole wave tests, while the other parameters were determined from triaxial tests. A back-calculation of the input triaxial tests is illustrated in figures 1a and 1b. It can be observed that, although the model describes well the evolution of the deviatoric stress tensor, it does not reproduce the decrease of dilatancy with the increase of the cell pressure.

Eo (100 kPa)	n	vo	ϕ (°)	ϕ_c (°)	a	b	c	α g
1000	0,5	0,35	38	35,5	0,0035	0	0	10

Table 1 : Set of parameters used for the prediction of the footing tests

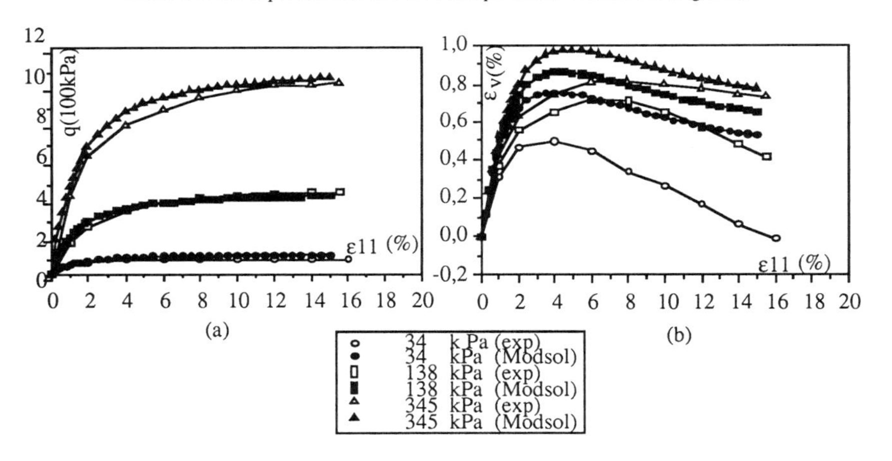

Figures 1 : Back calculation of input triaxial tests
a : deviatoric stress b : volumetric strain

PREDICTIONS OF THE FOOTING TESTS

Numerical simulations were carried out using the finite element program PECPLAS (Shahrour 1992). In order to simplify finite element calculation, predictions were performed assuming axisymetric conditions. A square footing (B x B) was replaced by a circular footing whose diameter is equal to $\sqrt{4/\pi}$ B.

The mesh used in the finite element calculation is illustrated in figure 2. It is composed of 124 eight-node elements.

Calculations were realized using the initial strain method. Convergence test was performed on the ratio (RC) of the unbalanced forces to that of the incremental applied load. A good convergence was obtained in the first steps (RC ≤ 0.001). In the last steps, a convergence of (RC ≤ 0.05) was obtained using small loading steps and a high number of iterations (up to 800).

The results of numerical predictions of four footing tests are illustrated in figure 3 togother with their bearing capacity as determined from the Terzaghi equation.

Loads corresponding to settlements of 25 and 150 mm are given in table 2. Since predictions were performed using an elastoplastic constitutive model, settlement due to creep was not computed.

	Footing 1 3 m x 3 m	Footing 2 1.5 m x1.5 m	Footing 3 3 m x 3 m	Footing 4 2m x 2m	Footing 5 1 m x 1m
Q_{25} (kN) settlement = 25 mm	1530	450		740	275
Q_{150} (kN) settlement = 150 mm	7560	2397		3720	887

Table 2 : Prediction of loads corresponding to settlements 25 and 150 mm

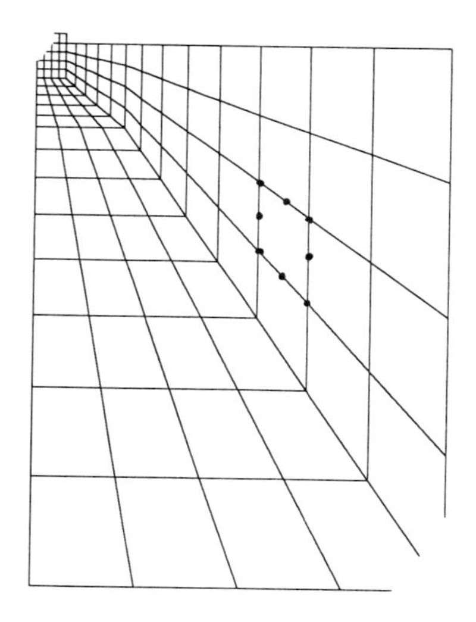

Figure 2 : Mesh used in the finite element calculation

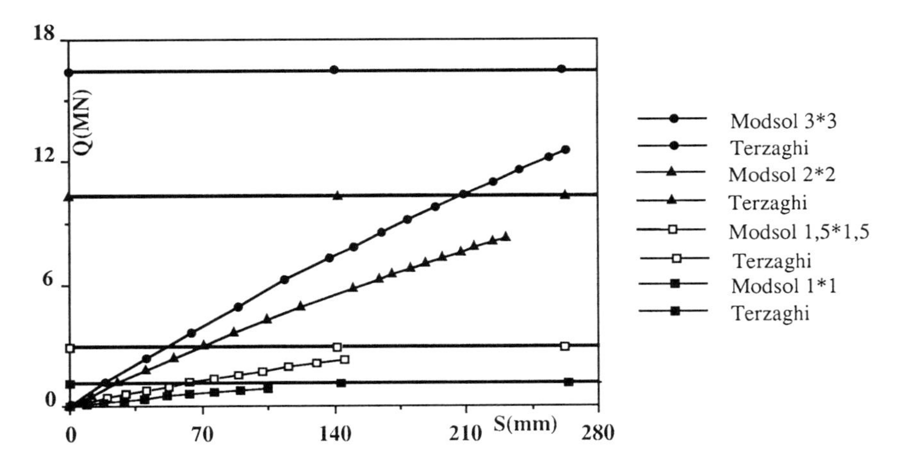

Figure 3 : Numerical predictions of the footing tests

REFERENCES

Luong M.P. "Etat caractéristique des sols" C.R Académie des Sciences, Paris, 287 B, pp. 305-307, 1978.

Schofield A.N. & Worth C.P. "Critical state soil mechanics" Mc Graw Hill Ed, London, 1968.

Shahrour I. "PECPLAS : A finite element software for the resolution of earthwork problems" Colloque International Géotechnique, Informatique, Edition Presse ENPC, Paris, pp. 327-334, Sept 1992.

Shahrour I. and Chehade W. "Development of a constitutive elastoplastic model for soils" XI International Congress on Rheology, Bruxelles, Edition P. Moldenaers & R. Keunigs, ELSEVIER, 1992.

Zienkiewicz O.C, Leung K.H, Pastor M. "Simple model for transient soil loading in earthquake analysis I. Basic model and its application" Int Jour. for Num. and Anal. Meth. in Geomechanics, Vol 9, pp. 453-476, 1985.

PREDICTION OF SHALLOW FOUNDATION SETTLEMENT ON SAND

By Mazen E. Adib[1], and Ramin Golesorkhi[2], Members, ASCE

Abstract: This paper summarizes the predicted settlements of five shallow spread-type footings with different sizes constructed on medium dense sand at the National Geotechnical Experimentation Site. The prediction is based mainly on in-situ SPT, CPT, and DMT results. Four different empirical or semi-empirical methods were used to arrive at the estimated settlements.

Introduction

The allowable bearing capacity of shallow spread-type footing on loose to medium dense sands is usually controlled by settlement rather than bearing capacity considerations. For routine-type design situations, The Standard Penetration Test results (SPT) are frequently used for this purpose, along with Peck, Hanson, and Thornburn (1974) empirical curves to estimate a conservative allowable bearing capacity for one-inch of total settlement. Recently, however, and as the Cone Penetrometer Tests (CPT) are becoming widely used, Schmertmann's semi-empirical method and its refinements (Fang, 1991) are being frequently used in practice.

Although available methods of calculating the settlements on clays have long considered explicitly the effect of overconsolidation, the methods for calculating the settlement on sands have lagged in this regard. Recent work by several researchers (Leonards and Frost, 1988, and Papadopoulos, 1992) has shown that the effect of overconsolidation in sands is very important, and should be considered if a reasonable estimate of the settlement is to be made. In this respect, the use of the Dilatometer Test (DMT) results, when combined with rational analytical methods, could provide more realistic and systematic estimates of the settlement of footings on sands.

Method of Estimating The Load-Settlement Relation

For the purpose of this prediction, we have used the following four different methods for estimating the load-settlement curves of the various foundations:

1. We used the SPT blow counts to estimate the allowable bearing pressure for one-inch of settlement from Peck, Hanson, and Thornburn (1974) empirical curves (herein referred to as PHT).
2. We also used the SPT blow counts to estimate the settlement of the footings using the method by Burland and Burbidge (1985) (herein referred to as SPT).
3. We used the results of the CPT with a method by Schmertmann (1978) (herein referred to as CPT), using $E_s = 2.5\, q_c$, where E_s is the equivalent Young's modulus for sand for footing-type loading, and q_c is the cone tip resistance.
4. We used the results of the CPT and the DMT with the method proposed by Leonards and Frost (1988) (herein referred to as DMT). In this method, the overconsolidation

---

[1] and [2] Senior Engineer and Project Engineer, respectively. Dames & Moore, 221 Main Street, Suite 600, San Francisco, CA 94105.

pressure is first estimated, as shown by Leonards and Frost. The equivalent Young's modulus in the overconsolidated portion of the loading is taken as 3.5 E_d, where E_d is the dilatometer modulus. The equivalent Young's modulus in the normally consolidated portion of the loading is taken as 0.7 E_d. The total settlement of the footing is the sum of the settlements in the overconsolidated and normally consolidated portions of the loading.

Results and Discussion

Load for 25 and 150 mm of settlement-The results of our analyses for the five footings are shown on Figures 1 through 5. In all of the above four methods, the in-situ test closest to the footing was selected. For footing 3, we used CPT-7 data to predict the settlement by Methods 3 and 4. CPT-7 data, however, were questionable at depth interval of 8 to 10 feet, where the tip resistances were close to zero, and the ratio of friction to tip resistance was zero. It is not known what caused these low values, but as shown on Figure 3, the predictions from Method 3 were significantly affected (we have used a value of 20 tons per square foot for E_s at depth interval between 8 and 10 feet to complete the prediction).

The analyses of the DMT data indicated that the sand is overconsolidated, with a calculated coefficient of lateral earth pressure at rest between 0.6 to 1. The overconsolidation ratios for this site were calculated in the range of 2 to 6.

Based on our evaluation of all four methods, we have selected our best estimate of the load on each footing at 25 and 150 mm of settlement, as summarized on Table 1. In most cases, and especially for the 25 mm settlement, we have relied heavily on the results of Method 4. For the 150 mm settlement, we took the average of Methods 3 and 4, except for footing 1, where we also considered the ultimate bearing capacity of the footing in accordance with Peck, Hanson and Thornburn (1974), and footing 2, where we used Method 3 only.

Creep Settlement between 1 minute and 30 minutes- Based on the pressuremeter test data of modulus versus time, we estimated that a reduction in the modulus of about 5 percent may occur between 1 and 30 minutes. Consequently, we estimated that the creep settlement is equal to about 5 percent of the total settlement at 1 minute.

Settlement in the year 2014- We do not believe that the settlement will increase appreciably in the next 30 years. and therefore, we have selected the same settlement as that after 30 minutes.

REFERENCES

Burland, J.B. and Burbide, M.C., Settlement of foundations on sand and gravel", Proceedings of the Institution of Civil Engineers, 78 Part 1, pp. 1325-1381, 1985.

Fang, H.Y., Foundation Engineering Handbook, Second Edition, Van Nostrand Reinhold, publishers, 1991.

Leonards, G.A. and Frost, J.D., Settlement of shallow foundations on granular soils, Journal of Geotechnical Engineering, ASCE, 114, No.7, pp. 791-809, 1988.

Papadopoulous, B.P., Settlements of shallow foundations on Cohesionless soils, Journal of Geotechnical Engineering, ASCE, 118, No.3, pp. 377-393, 1992.

Peck, R.B., Hanson, W.E., and Thornburn, T.H., Foundation Engineering, Second Edition, John Wiley and Sons, publishers, 1974.

Schmertmann, J.H., Guidelines for cone penetration test performance, and design, Federal Highway Administration, Report FHWA-TS-78-209, 1978.

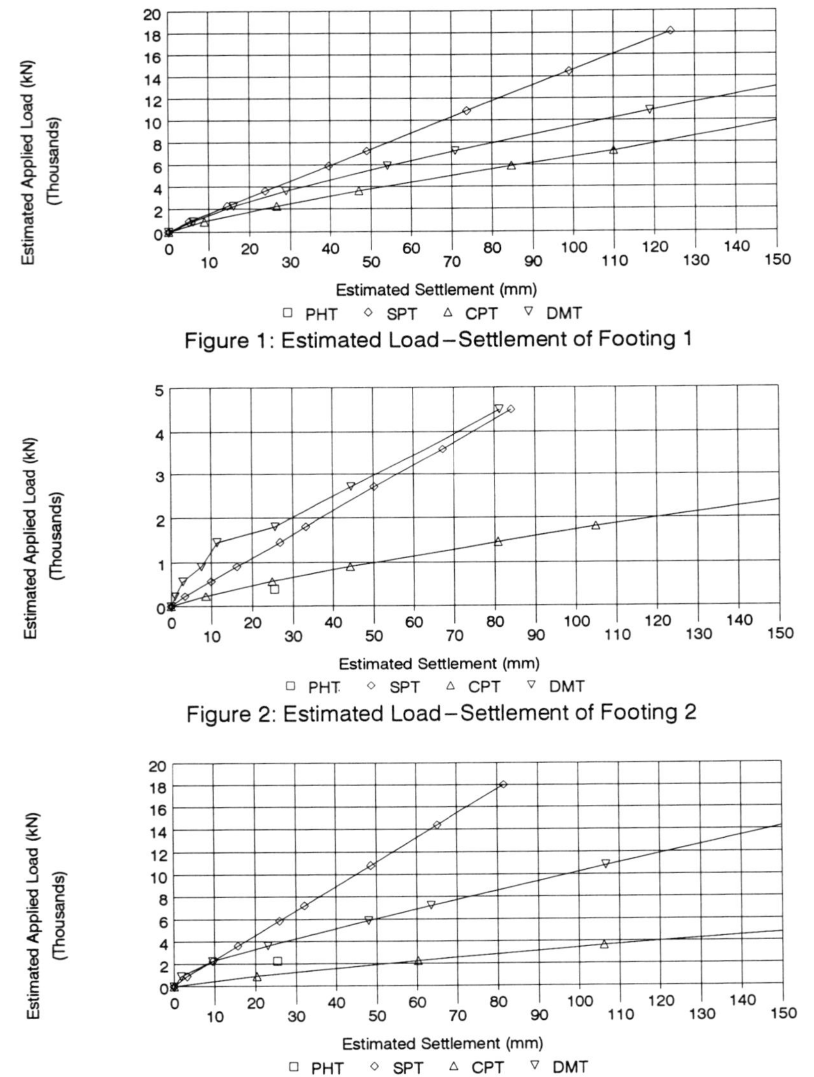

Figure 1: Estimated Load–Settlement of Footing 1

Figure 2: Estimated Load–Settlement of Footing 2

Figure 3: Estimated Load–Settlement of Footing 3

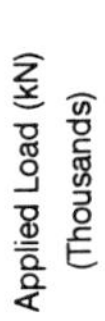

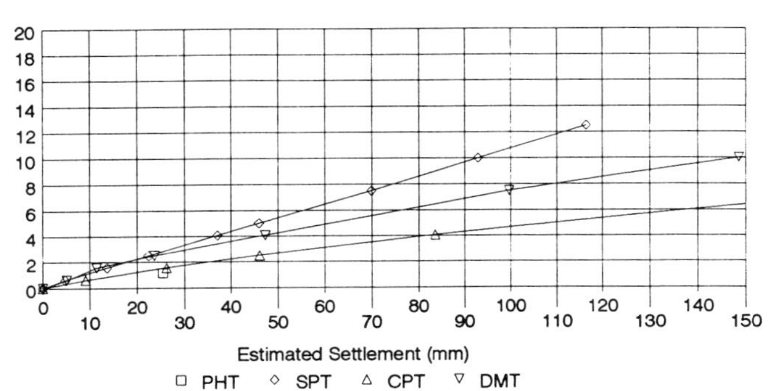

Figure 4: Estimated Load – Settlement of Footing 4

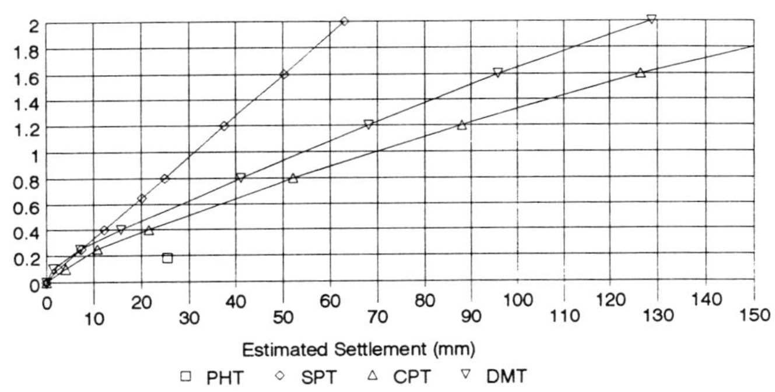

Figure 5: Estimated Load – Settlement of Footing 5

TABLE 1: SUMMARY OF PREDICTIONS FOR THE FIVE FOOTINGS

	Footing 1 3mx3m	Footing 2 1.5mx1.5m	Footing 3 3mx3m	Footing 4 2.5mx2.5m	Footing 5 1mx1m
Load for 25 mm of settlement Q25 on the 30 minute load settlement curve (kN)	3300	1000	3600	2600	550
Load for 150 mm of settlement Q150 on the 30 minute load settlement curve (kN)	11500	3000	12000	8000	1600
Creep Settlement Between 1 minute and 30 minute for Q25 (mm)	1.5		1.5		
Settlement in the year 2014 under Q25 (mm)	25		25		

PREDICTED LOAD-SETTLEMENT BEHAVIOR OF SPREAD FOOTINGS ON SAND

by

Geotechnical Engineering Division[1]
Utah State University
Logan, Utah 84322-4110
U. S. A.

INTRODUCTION

This paper presents the methodology that was adopted to predict the settlements of spread footings on sand in response to the prediction request for the spread footing prediction event sponsored by the Federal Highway Administration. This event is held at the occasion of the ASCE Specialty Conference on Vertical and Horizontal Deformations of Foundations and Embankments (SETTLEMENT '94) to be held at Texas A&M University on June 1994. The primary goal of the event is to evaluate the capability of the profession to predict the behavior of spread footings on sand at small and large deflections. To achieve the goal, five square footings were constructed at the site with dimensions ranging from 1 m by 1 m to 3 m by 3 m. All the footings are planned to be load tested to 0.15 m of settlements.

The prediction packet that was provided to the interested participants contains various sets of geotechnical information of the site based on different in situ and laboratory tests that were conducted prior to the construction of the footings. The predicted values reported in this paper were derived from three sets of data that were included in the prediction packet. These data are the standard penetration test (SPT) blow counts, deformation moduli from pressuremeter tests, and stress-strain data from the triaxial tests. These three sets of data were chosen so that an independent check of settlements calculated from both empirical and theoretical approaches can be made. Furthermore, all the predictions discussed in this paper were made based on isolated footing analyses, after preliminary stress distribution analyses had shown that interaction between footings can be neglected.

LOAD-SETTLEMENT EVALUATION PROCEDURE

The methods currently used to estimate settlements of spread footings on sand can be generally grouped into two categories based on the theories on which they were derived. These theories are the bearing capacity theory and the theory of elasticity. The former category usually requires the standard penetration test blow count as the fundamental soil property, while the latter mainly depends on the deformation modulus of the soil. Both methods have been used successfully in the past but have the limitation that settlement is assumed to vary linearly with the applied load. This assumption may be valid at low strain levels and was used in this paper to predict the loads associated with 25 mm of settlement. On the other hand, loading the footings up to a settlement of 150 mm will cause strains that are beyond the elastic range and may approach failure as the ultimate bearing capacity is exceeded. Thus, a nonlinear finite element analyses was carried out to predict loads at large deformations. The finite element solution was validated by comparing the predicted loads at 25 mm settlements to those obtained by the other calculation procedures used in this paper.

---

[1] Refer to the last page of this paper for complete list of authors.

SPT-Based Methods

Most of the hand calculation methods published in the literature are based on SPT blow counts. To establish a possible range of predicted loads at 25 mm using an SPT-based solution, the settlement studies made by Tan & Duncan (1991) was used as a guideline. From the 76 cases of spread footings that were studied, they concluded that the methods of Alpan (1964), Schultz & Sherif (1973), Parry (1971), D'Appolonia & D'Appolonia (1970), Peck & Bazaraa (1969) and U.S. Navy (1982) were among those which, on the average, can yield the highest accuracy of predicted settlements. These six methods were used in this paper with the assumption that the calculated settlements correspond to the 30-minute load-settlement curves.

Elasticity-PMT Approach

This method applies the results of pressuremeter (PMT) tests to calculate settlements using the elasticity formula (Janbu et al, 1956):

$$s=\mu_o\mu_1(1-\nu^2)\frac{qB}{E} \quad (1)$$

where s is the footing settlement, μ_o and μ_1 are influence factors, ν is the Poisson's ratio which was assumed to be 0.33, q is the bearing pressure, B is the footing width, and E is the pressuremeter modulus. The values of E used in Eq. 1 were the initial deformation modulus (E_o) derived from pressuremeter test data and was averaged over the depth B below the footing base following the averaging technique proposed by Schmertmann (1970). These values were then multiplied by a factor of 2 as suggested by Briaud (1992). Furthermore, the values of E_o given in the prediction packet were assumed to correspond to the 1 minute values, E_1. To estimate the 30-minute deformation modulus, E_{30}, the following expression was used:

$$\frac{E_t}{E_1}=\left(\frac{t}{t_1}\right)^{-n_c} \quad (2)$$

where E_t is the deformation modulus at any time t (minutes), t_1 is the time at which E_1 was measured (i.e. 1 minute) and n_c is the time exponent obtained as the slope of regression lines for the normalized time-dependent plots of pressuremeter modulus data.

Nonlinear Finite Element Analysis

A nonlinear finite element analysis was used to estimate the loads causing a 150 mm settlement. Each footing was analyzed independently using a finite element mesh consisting of 399 nodal points and 358 plane strain soil elements. The footings were modeled as beam elements that have both axial and flexural stiffnesses. The loads were incrementally applied as distributed loads at the base of the footings.

The soil elements were modeled by a hyperbolic stress-strain law (Duncan & Chang, 1970) with parameters estimated from the triaxial test data. The parameters used in the finite element analysis are as follows: unit weight (γ) = 18.0 kN/m^3, Young's modulus number (K) = 220, Young's modulus exponent (n) = 0.84, failure ratio (R_f) = 0.89, bulk modulus number (K_b) = 200, bulk modulus exponent (m) = 0.42, cohesion (c) = 0.0, friction angle (ϕ_o) = 36°, reduction in friction angle for 10-fold increase in pressure ($\Delta\phi$) = 0.0, and unload-reload Young's modulus (K_{ur}) = 330.

The soil model used, being based on generalized Hooke's law, is limited in accuracy for stress levels in excess of about 80%. Since the footing stresses corresponding to 150 mm of settlement are believed to cause plastic deformations, it is believed that plasticity models would be more appropriate to model the soil behavior at large deformations. To account for this limitation, the results of finite element analysis were interpreted based on judgment from the knowledge of the estimated ultimate bearing capacities of the footings. This was further supported by the anticipated shape of the load-settlement curves based on footing widths as suggested by Terzaghi & Peck (1967).

Long Term (Creep) Settlements

The creep settlements that occur between a 1 minute and 30 minute period were estimated using Eqs. 1 and 2. For the 20-year settlements (s_{2014}), the procedure suggested by Schmertmann (1970) using the following equation was used:

$$s_t = s_o\left(1 + 0.2\log\frac{t}{t_o}\right) \tag{3}$$

where s_o is the settlement at time t_o (equal to 0.1 year) and s_t is the settlement at any time t (years). The value of s_o was estimated using Eqs. 1 and 2.

PREDICTED RANGES AND BEST ESTIMATES

The predicted range of loads to cause 25 mm of settlements (Q_{25}) for the five footings are given in Table 1. The best estimates were taken as the mean of the values predicted by the eight methods, except for footing 5 wherein the calculated mean exceeds the ultimate bearing capacity. In such a case, the latter value was adopted. The best estimates are shown in Table 2. Table 1 also shows the probable range of loads to cause 150 mm of settlements. The lower limits for each range are the lower bound estimates of the ultimate bearing capacity of the footings using the N_γ values of Terzaghi (1943). The upper limits are the magnitudes predicted by the finite element analysis. The best estimates of Q_{150} shown in Table 2 are the ultimate bearing capacities predicted from the values of N_γ based on the limit analysis solution by Chen (1975). These values are believed to be more reasonable than simply picking the average of the lower and upper limits.

The probable range of creep settlements that can be expected between the 1-minute and 30-minute period for Q_{25} is given in Table 2. These values were estimated from the time-dependent pressuremeter modulus data. Using Eq. 3, the settlement after 20 years due to Q_{25} are also given in Table 2.

REFERENCES

Alpan, I. (1964). "Estimating the Settlements of Foundations on Sands," Civil Eng'g. and Public Works Review, pp. 1415-1418.

Briaud, J.L. (1992). The Pressuremeter. A.A. Balkema, Rotterdam.

Chen, W.F. (1975). Limit Analysis and Soil Plasticity. Elsevier, Amsterdam.

D'Appolonia, D.J. & D'Appolonia, E. (1970). "Closure to Settlement of Spread Footings on Sand," J. of Soil Mech. & Found. Div., ASCE, Vol. 96, No. SM2, pp. 754-762.

Duncan, J.M. & Chang, C.Y. (1970). "Nonlinear Analysis of Stress and Strains in Soils," J. of Soil Mech. & Found. Div., ASCE, Vol. 96, No. SM5.

Janbu, N., Bjerrum, L. & Kjaernsli, B. (1956). "Guidance in the Solution of Foundation Problems," Norwegian Geotechnical Institute Publication No. 16.

Parry, R.H.G. (1971). "A Direct Method of Estimating Settlements in Sand from SPT Values," Proc. Symp. on the Interaction of Structure and Foundation, Birmingham, pp. 29-32.

Peck, R.B. & Bazaraa, A.R.S. (1969). "Discussion to Settlement of Spread Footings on Sand," J. of Soil Mech. & Found. Div., ASCE, Vol. 95, No. SM3, pp. 905-909.

Schultz, E. & Sherif, G. (1973). "Prediction of Settlements from Evaluated Settlement Observations for Sands," Proc. 8th ICSMFE, Moscow, Vol. 1.3, pp. 225-230.

Schmertmann, J.H. (1970). "Static Cone to Compute Static Settlement Over Sand," J. of Soil Mech. & Found. Div., ASCE, Vol. 96, No. SM3, pp. 1011-1043.

Tan, C.K. & Duncan, J.M. (1991). "Settlement of Footings on Sands - Accuracy and Reliability," Proc. Geotech. Eng'g. Congress, ASCE Spec. Pub. No. 27, Vol. 1, McClean, F.G., Campbell, D.A. and Harris, D.W. (eds.), Boulder.

Terzaghi, K. (1943). Theoretical Soil Mechanics. John Wiley & Sons, New York.
Terzaghi, K. & Peck, R.B. (1967). Soil Mechanics in Engineering Practice. 2nd Edition, John Wiley & Sons, Inc., New York, 529 p.
U.S. Dept. of Navy (1982). "Soil Mechanics-Design Manual 7.1," U.S. Govt. Printing Office, Washington, D.C., 348 p.

TABLE 1. Predicted Range of Loads to Cause 25 mm and 150 mm of Settlements

	Footing 1 3m x 3m	Footing 2 1.5m x 1.5m	Footing 3 3m x 3m	Footing 4 2.5m x 2.5m	Footing 5 1m x 1m
Range of Q_{25} (kN)	3340 to 6880	1140 to 2112	3419 to 7013	2367 to 3883	667 to 1416
Range of Q_{150} (kN)	12219 to 20745	1946 to 3226	13396 to 20960	6571 to 13092	838 to 1690

TABLE 2. Prediction Summary of the Required Quantities for the Five Spread Footings

	Footing 1 3m x 3m	Footing 2 1.5m x 1.5m	Footing 3 3m x 3m	Footing 4 2.5m x 2.5m	Footing 5 1m x 1m
Q_{25} (kN)	4668	1644	4808	3087	838
Q_{150} (kN)	13943	2119	15143	7375	900
Creep Δs_1 (mm)	3.0	(not required)	3.0	(not required)	(not required)
s_{2014} (mm)	40.54	(not required)	40.58	(not required)	(not required)

COMPLETE LIST OF AUTHORS

The following are the authors of this prediction paper:

Casan L. Sampaco, Loren R. Anderson, Joseph A. Caliendo, Abdel Agallouch, Hogan Chang, Lin Chia-Ching, Elhassan Elhassan, Kyle Gorder, Roger Greaves, Makarand S. Jakate, Won C. Kim, Arthur L. Moss, and Darrin Sjoblom.

PREDICTING THE LOAD-DISPLACEMENT BEHAVIOUR OF SPREAD FOOTINGS ON SAND

Guido Gottardi[1] and Paolo Simonini[2]

Predicting settlements of shallow footings on sand in a reliable and consistent way is still a formidable task of modern geotechnical engineering, even if in the last 30 years a large number of methods has been proposed; most of them provide a relationship between soil characteristics - measured in situ by penetration tests - and either "elastic" soil moduli or, alternatively, observed displacements of existing structures. However loading values requested by the prediction event are related to unusual settlements ($s/B = 0.8\% \div 2.5\%$ with $s = 25$ mm and $s/B = 5\% \div 15\%$ with $s = 150$ mm) and a different procedure, which takes into account the overall load-settlement behaviour, has to be used. Since parameter evaluation is unfortunately affected by a high degree of approximation a simple method has been adopted, which starts from an exponential relationship (Butterfield, 1980) between the load Q and the settlement s:

$$Q = Q_{max}\left[1 - \exp\left(-\frac{Ks}{Q_{max}}\right)\right], \tag{1}$$

where Q_{max} is the vertical-central ultimate load of the footing and K represents the initial stiffness of the load-settlement curve. Eq. (1) has proved to fit quite well a large amount of experimental data from plate loading tests on sand (Gottardi, 1992), particularly in the case of a loose/medium-dense sand whose strain-hardening behaviour is well represented by a horizontal final slope. Therefore this approach enables us to describe the load-displacement behaviour of spread footings resting on a granular material through the evaluation of only two parameters, namely the ultimate load and the initial stiffness.

Q_{max} can be determined by means of the classical superposition bearing capacity formula for a square footing of breadth B:

$$Q_{max} = B^2\left(0.5\gamma_1 B N_\gamma s_\gamma + \gamma_2 D N_q s_q\right), \tag{2}$$

[1] Research Fellow, Istituto di Costruzioni Marittime e di Geotecnica, University of Padova, Via Ognissanti 39, 35129 Padova, ITALY.
[2] Researcher, Istituto di Costruzioni Marittime e di Geotecnica, University of Padova, Via Ognissanti 39, 35129 Padova, ITALY.

where D is the embedment depth, N_γ,N_q and s_γ,s_q are respectively the bearing capacity and the shape factors. The following values have been adopted: γ_1 = 15.65 kN/m^3 and γ_2 = 15.28 kN/m^3; s_γ = 1-0.4B/L = 0.6 and s_q = 1+0.2B/L = 1.2. N_γ and N_q are functions of the angle of shearing strength ϕ'; from triaxial compression tests an average value of ϕ' = 35° for the all site can be deduced. However, eq. (2) has been derived for strip footings and then - in order to obtain an ultimate load prediction not too conservative - a ϕ'-value determined in plain strain conditions should be used. Following Wroth (1984), ϕ' = 38° have been eventually selected. N_q factor can be therefore calculated using the expression $N_q = e^{\pi \tan\phi'}(1+\sin\phi')/(1-\sin\phi') = 48.9$ whereas N_γ = 102.1 has been adopted as proposed by Chen (1975) through the limit analysis. Table 1 shows the ultimate loads calculated with eq. (2) following the "as-built" dimensions of the five spread footings. The high values indicated in the last column suggest that the ultimate load of the largest footings will not be probably reached.

TABLE 1

Footing	Length x Width (m)	Thickness (m)	Embedment Depth (m)	Q_{max} (kN)
1	3.004 x 3.004	1.219	0.762	19167
2	1.505 x 1.492	1.219	0.762	3150
3	3.023 x 3.016	1.346	0.889	20484
4	2.489 x 2.496	1.219	0.762	11678
5	0.991 x 0.991	1.168	0.711	1093

From classical elastic theory the initial load-settlement ratio K can be expressed as a function of the footing breadth and the soil Young modulus E':

$$K = \frac{Q}{s} = \frac{BE'}{I_s} \qquad (3)$$

The influence factor I_s depends on the shape and rigidity of the foundation, the Poisson's ratio ν' and the ratio of the depth of the compressible stratum to the foundation breadth; a value I_s = 0.69 should well fit our case. The problem is so led back to the suitable determination of E'.
The selection of an unique value of E' to insert in eq. (3) is crucial, but extremely difficult because the "equivalent elastic" modulus depends - for a given soil - on several factors like the relative density, the level of the mean effective stress and of the average vertical strain and the stress history of the deposit.

In our case data from many in situ (SPT, CPT, PMT, DMT and CHT) and laboratory (RCT) tests are provided. An examination of CPT profiles and especially SPT ones (Fig. 1) - essentially constant with depth - would suggest that a lightly overconsolidated sand must be considered. Furthermore, still looking at Fig. 1, no significant differences can be appreciated among the six SPT logs. Therefore, in this particular case, the effects of the increase of the geostatic effective stress and the spatial variability can be neglected and a constant value of E' with depth and below the five footings has been selected.
Most of the available tests would allow for the determination of an "elastic" modulus, although related to very different strain levels. The required initial stiffness K can be referred to a rather low strain level, i.e. less than 0.1%. Therefore, on the right side of Fig. 2 the elastic moduli determined from RCT (E_{max}), PMT (E_R) and CHT are first compared. Values from RCT and

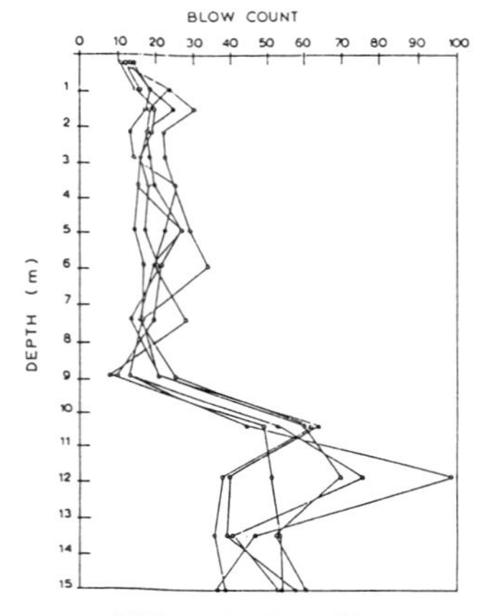

FIG. 1 SPT profiles

CHT have been calculated through the elastic relationship $E_{max} = 2G_{max}(1+\nu')$ with $\nu' = 0.20$ and taking into consideration the level of the in-situ vertical effective stress below the centre of each footing (left side of Fig. 2). The wide scatter of data shown can be explained again with the different strain level characteristic of each test.

In addition SPT, CPT and DMT results have been examined, although the large deformations caused by the penetration of such devices mostly obliterate the effects of stress and strain history in the soil. Relationships between penetration parameters (N_{SPT}, q_c and E_D) and "elastic" moduli taking into account the marked effect of the stress history should then be used. The back-analysis of several field data of shallow footings on sand performed by Berardi et al. (1991) enables to provide some operational charts in which existing cohesionless deposits are grouped following their stress history and age. As our sand deposit can be assumed lightly over-consolidated and probably aged, an elastic modulus in the range 50÷70 MPa results from this procedure.

On the basis of the above considerations an average elastic modulus E' = 60 MPa has been eventually selected. Introducing this value in eq. (3) and the resulting K in eq. (1), together with Q_{max} values from Table 1, exponential curves of Fig. 3 can be drawn. Loading values corresponding to s = 25 mm and s = 150 mm are reported in Table 2. Note that, for footings 2 and 5, s = 150 mm should practically be related to their ultimate load.

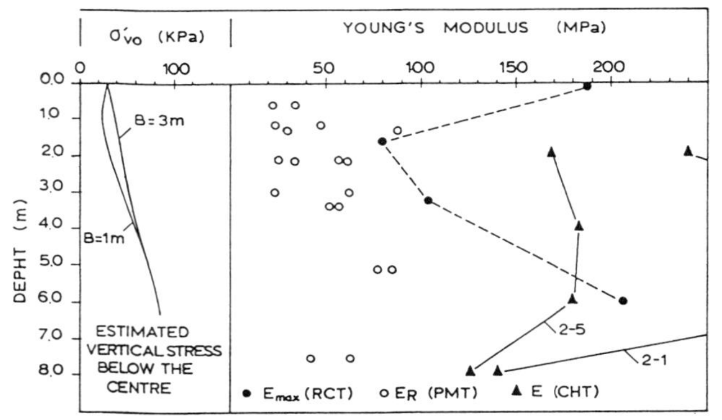

FIG 2 Elastic moduli from RCT, PMT and CHT tests

The evaluation of time-dependent settlements under constant load Q_{25} for the B=3m-footings has been performed using the pressuremeter test results. The equation fitting the experimental data provided

$$\log_{10}\left(\frac{E'}{E'_{1min}}\right) = -n_c \log_{10}\left(\frac{t}{t_{1min}}\right) \tag{4}$$

shows a reduction of the pressuremeter modulus as a function of time. An average value of the creep exponent n_c = 0.0135 can be assumed. Since our prediction is based on an exponential curve whose initial stiffness depends on the equivalent elastic modulus, a load-settlement curve corresponding to t = 1 min has been determined by introducing in eqs. (3) and (1) an increased value of E' deduced from eq. (4). The creep settlement has been then calculated as a difference between settlements for t = 1 min and t = 30 min. The same considerations have been extended to the calculation of the settlement in the year 2014. Creep settlements are reported in Table 2.

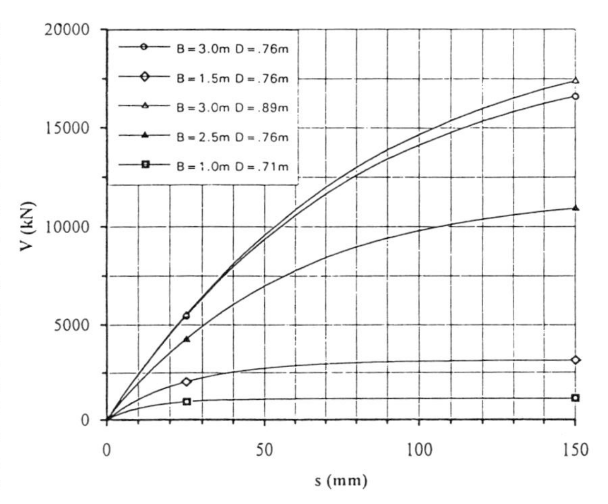

FIG. 3 Predicted load-displacements curves

TABLE 2

	Footing 1 3m x 3m	Footing 2 1.5m x 1.5m	Footing 3 3m x 3m	Footing 4 2m x 2m	Footing 5 1m x 1m
Load for 25 mm settlement Q_{25} on the 30 min load settlement curve (kN)	5446	2008	5526	4271	935
Load for 25 mm settlement Q_{25} on the 30 min load settlement curve (kN)	16587	3143	17379	10918	1093
Creep settlement between 1 and 30 min for Q_{25} (mm)	1.1	–	1.1	–	–
Settlement in the year 2014 under Q_{25} (mm)	29.7	–	29.7	–	–

REFERENCES

Berardi, R., Jamiolkowski, M. and Lancellotta, R. (1991). "Settlement of shallow foundations in sands. Selection of stiffness on the basis of penetration resistance." *Proc. Geotech. Eng. Congress*, GT/Div. ASCE, Boulder, Colorado, pp. 185-200.

Butterfield, R. (1980). "A simple analysis of the load capacity of rigid footings on granular materials." *Journée de Géotechnique*, ENTPE, Vaulx-en-Velin, pp. 128-137.

Chen, W.F. (1975). "*Limit analysis and soil plasticity.*" Elsevier, Amsterdam.

Gottardi, G. (1992). "Modellazione del comportamento di fondazioni superficiali su sabbia soggette a diverse condizioni di carico." Ph.D. Thesis, University of Padova.

Wroth, C.P. (1984). "The interpretation of in situ soil tests." *Géotechnique*, 34, pp. 449-489.

SETTLEMENT OF TEST FOOTINGS:
PREDICTIONS FROM THE UNIVERSITY OF NEW MEXICO

Koon Meng Chua[1] M.ASCE, Ling Xu[2], Eric Pease[2], and Sameer Tamare[2]

INTRODUCTORY SUMMARY

An adequately designed foundation will enable the soil to carry loads without failing in shear and with an acceptable settlement. The first requirement concerns a limiting equilibrium condition and can be determined using bearing capacity equations. The second requirement is based on a subgrade modulus which can be either determined in situ using a bearing plate test or numerically using soil properties inferred from other tests. This paper presents the predictions, made by the New Mexico team, of the load-settlement characteristics of five (5) footings and the long-term settlement characteristics of a 3 m x 3 m footing at the National Science Foundation, Texas A&M University Test Site. The finite element approach was used to make the required predictions. Soil properties were based on information found in Gibbens and Briaud (1993). The predictions obtained are presented in tables and figures that are self-explanatory. FIG. 1 illustrates the footings and the site conditions.

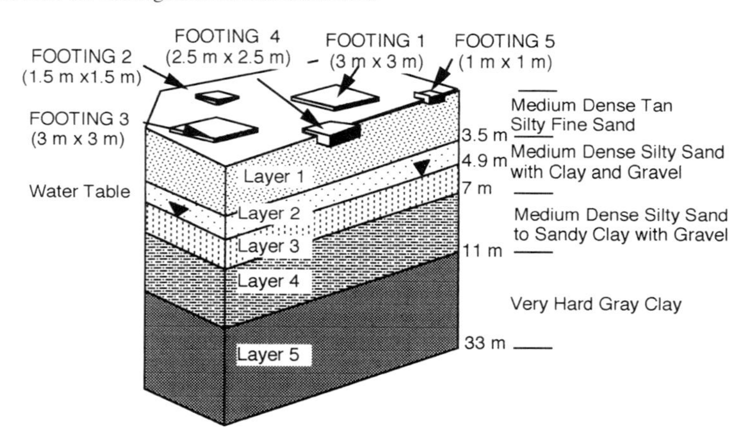

FIG 1. Soil and Site Conditions

THE APPROACH: USING THE FINITE ELEMENT METHOD

GEOT2D which was developed at the University of New Mexico was used. This is an updated Lagrangian code which allows large deformations in the soil structure system. The footing tests were modeled using as axisymmetrical solids of revolution. The dimension of the footings was sized to provide the same bearing area (i.e., using equivalent radii). The concrete material and soils were modeled using iso-parametric continuum elements. Interface elements were used

[1]Asst.Prof., Dept. of Civil Engrg., Univ.of New Mexico, Albuquerque, NM 87131
[2]Grad.Students, Dept. of Civil Engrg., Univ.of New Mexico, Albuquerque, NM 87131

between the footing and the soil. Slippage, if any, is assumed to occur in the soil. The derivation of this interface element and its form for use in the axisymmetric solid of revolution is extensively discussed in Yuan and Chua (1992). The soils were modeled using hyperbolic stress-strain curves (Duncan et al., 1980). Table 1 shows the soil parameters used in the simulation. These values were either based on laboratory tests, inferred from field testing, or obtained from a method developed from a soil database (Chua, 1986). The symbols used for describing the stress-strain equation are consistent with that found in Duncan et al. (1980).

The soil was modeled as a linear viscoelastic material when determining settlement histories. This approach was extensively discussed in Chua and Lytton (1987 and 1989). It was shown that under certain conditions, such as those encountered here, the relaxation moduli of the different soils can be represented by a simplified "power law",

$$E(t) = E_1 \, t^{-m}$$

where

E_∞ = the relaxation modulus of the material at a very long time, and

E_1 = the intercept and m is the rate of degradation or relaxation of the material.

The creep rate for silty and sandy soils was discussed in Chua and Lytton (1986).

Primary consolidation was not considered since the stress level at the hard clay layer was very low.

FIG. 2 shows the undeformed and the deformed mesh with the principal stress vectors for Footing #1.

TABLE 1. Summary of Soil Properties Used in Simulations

Layer No.	Soil Classification	Depth (m)	Depth (ft)	Unit Weight (pcf)	Strength Parameters: Friction Angle ϕ, (deg)	Strength Parameters: Cohesion c, (psi)	Stress-Strain Parameters: Modulus Number K	Stress-Strain Parameters: Modulus Exponent n	Stress-Strain Parameters: Failure Ratio R_f	Stress-Strain Parameters: Poisson's Ratio, ν	Creep Power Law Exponent, m
1	SP	0.6	2	110[a,d]	35.3[c]	3[c]	404.2[a,d]	-0.45[a,d]	1.0	0.3	0.0159[b]
	SP-SM	3.0	9.8				268.6	0.70			
2	SP(?)	3.6	11.8	110	33.5[a,d]	20[a,d]	300	0.24	1.0	0.3	0.0162
		4.2	13.8				270	0.28			
		4.8	15.8				270	0.28			
3	SP(?)	5.1	16.7	47.6	35.7	17.1	340	0.18	1.0	0.3	0.0206
		6.0	19.7				360	0.14			
		6.9	22.6				335	0.17			
4	SP(?)	7.1	23.3	47.6	33.4	12.6	335	0.17	1.0	0.3	0.044
		8.0	26.3				275	0.23			
		9.0	29.5				260	0.22			
		10.0	32.8				260	0.22			
		10.9	35.8				260	0.22			
5	CL(?)	11.1	36.4	47.6	1.0	36.7	288.9	0.27	1.0	0.4	0.068
		12.0	39.4				386.9	0.10			
		13.0	42.7				530.6	-0.10			
		15.0	49.2				488.9	-0.05			
		>15.0	49.2				488.9	-0.05			

[a]Standard Penetration Test [b]Pressuremeter Test Data
[c]Laboratory Test [d]Estimated
[a,d]Chua's Method from Soils Database

PREDICTIONS

The long-term settlement history for a 3m x 3m footing is shown in FIG. 3. This prediction applies to both Footing #1 and Footing #3. The load-settlement curves for the five footings are shown in FIG. 4. TABLE 2 summarizes the predictions required in this exercise.

TABLE 2 Summary of Predictions

	Footing 1 3 m x 3 m	Footing 2 1.5 mx 1.5 m	Footing 3 3 m x 3 m	Footing 4 2.5 m x 2.5 m	Footing 5 1 m x 1 m
Load for 25 mm of settlement Q25 on the 30 minute loading settlement curve (kN)	1455.6 kN (327.2 kips)	539.6 kN (121.3 kips)	1452.4 kN (326.5 kips)	1008.6 kN (249.2 kips)	313.3 kN (70.4 kips)
Load for 150 mm of settlement Q150 on the 30 minute loading settlement curve (kN)	3344,8 kN (751.9 kips)	937.1 kN (210.7 kips)	3320.1 kN (746.4 kips)	2413.0 kN (542.5 kips)	500.6 kN (112.5 kips)
Creep settlement between 1 minute and 30 minutes for Q25,Ds (mm)	5.8 mm (0.23")		5.8 mm (0.23")		
Settlement in the year 2014 under Q25 (mm)	186.9 mm (7.34")		187.3 mm (7.37")		

REFERENCES

Chua, K.M. & Lytton, R.L. (1987). A Method of Time-Dependent Analysis Using Elastic Solutions for Non-Linear Material. International Journal for Numerical and Analytical Methods in Geomechanics, Vol.11 No.4, John-Wiley & Sons, July - August, 1987, pp.421-431.

Chua, K.M. and Robert L. Lytton, R.L. (1989). Viscoelastic Approach to Modeling the Performance of Buried Pipes. ASCE, Journal of Transportation Engineering, Vol.115, No.3, May, 1989, pp.253-269.

Chua, K.M. (1986). Time-Dependent Interaction of Soil and Flexible Pipe. Ph.D. Dissertation, Texas A&M University, May, Appendices A & B.

Chua, K.M. and Robert L. Lytton, R.L. (1986). Time-Dependent Properties of Embedment Soil Back-Calculated from Deflections of Buried Pipes. Institution of Engineers, Specialty Geomechanics Symposium, "Interpretation of Field Testing for Design Parameters", Adelaide, Australia, August.

Duncan, J.M.; Bryne P.; Wong, K.S.; and Mabry, P. (1980). Strength, Stress-Strain and Bulk Modulus Parameters for Finite Element Analysis if Stresses and Movement in Soil Masses. report No. UCB/GT/80-01, University of California, Berkeley, Vol.I, pp. 19-25.

Gibbens, R. and Briaud, J.-L. (1993). Data and Prediction Request for the Spread Footing Prediction Event Sponsored by FHWA at the Occasion of the ASCE Specialty Conference: Settlement '94, at Texas A&M University during June 16 - 18 of 1994.

Yuan Z. & Chua, K.M. (1992). Exact Formulation of an Axisymmetric Interface Element Stiffness Matrix. J. Geotechnical Engineering, Vol. 118 No.8, August, p. 1264- 1272.

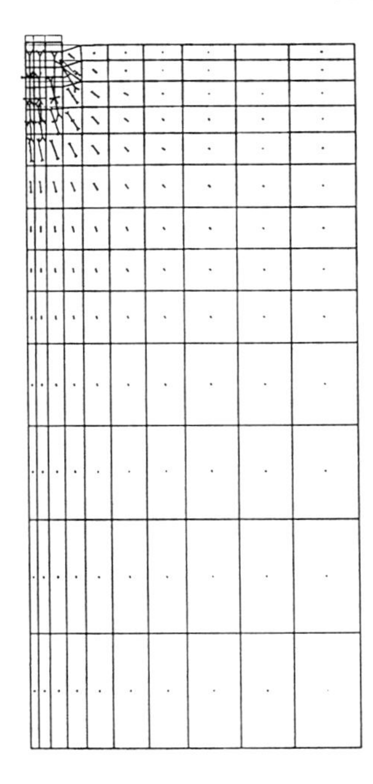

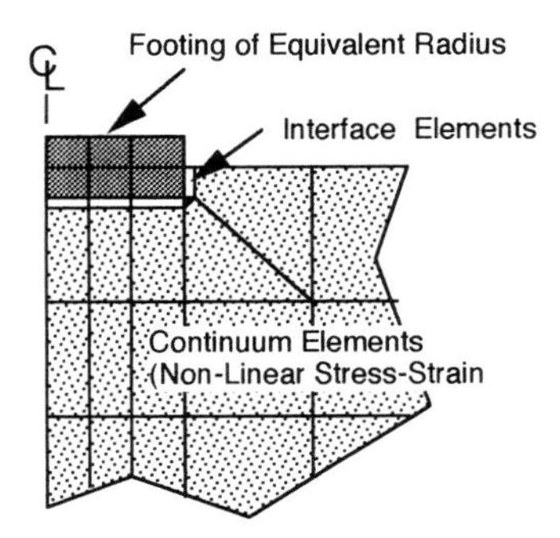

FIG. 2 Finite Element Simulation of a Footing

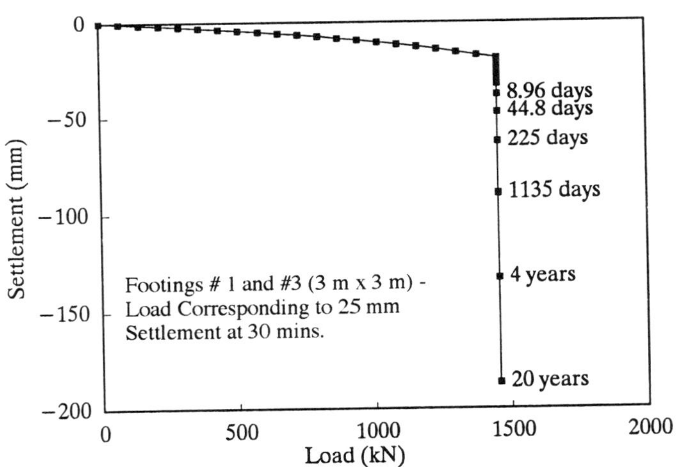

FIG. 3 Predicted Long-Term Settlement Curve for the 3 m x 3 m Footing

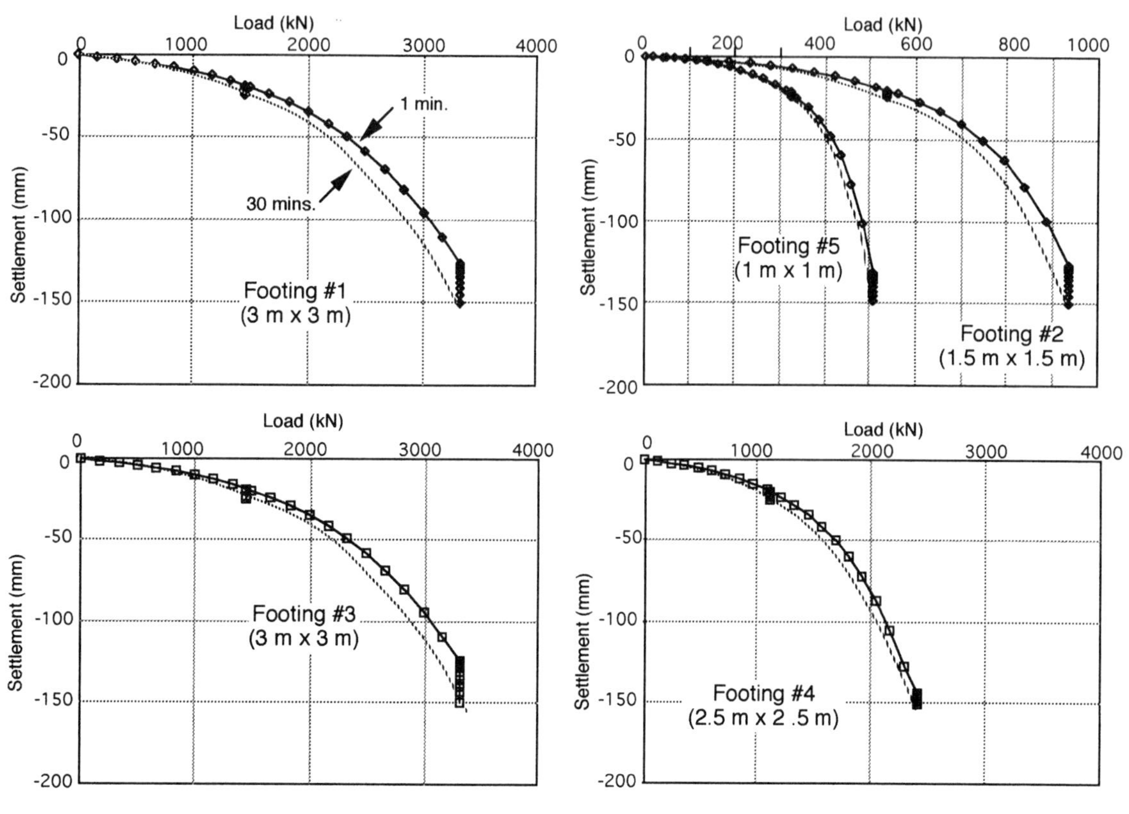

FIG. 4 Predicted Load-Settlement Curves for Footings for 1 to 30 mins. Loading Time

BEHAVIOR PREDICTION FOR SPREAD FOOTINGS ON SAND

By Sujit K. Bhowmik[1]

ABSTRACT: Predictions of the load-settlement behavior of five spread footings on sand are presented in this paper. The footings, having dimensions ranging from 1m x 1m to 3m x 3m, were constructed and load-tested as part of a prediction symposium held at Texas A&M University during the ASCE specialty conference SETTLEMENT '94. Before footing construction, field and laboratory soil investigations were conducted to obtain soil behavioral information and the results were provided to those interested in making a prediction of the load-settlement behavior of the footings. The predictions presented herein are based on the results of Standard Penetration Tests, Electric Piezocone Penetration Tests and Pressuremeter Tests performed at the site. The predictions include loads on the five footings corresponding to given settlements at 30 minutes after load application, creep settlements of two footings under given loads during the period from 1 minute to 30 minutes after load application, and settlements of two footings under given loads at 20 years after load application.

INTRODUCTION

At the occasion of the ASCE specialty conference SETTLEMENT '94 held at Texas A&M University on June 16, 17 and 18, 1994, the Federal Highway Administration sponsored a symposium to evaluate the ability of the geotechnical profession to predict the load-settlement behavior of spread footings on sand. Five square footings, designated as Footing 1 through Footing 5 and having dimensions of 3m x 3m, 1.5m x 1.5m, 3m x 3m, 2.5m x 2.5m and 1m x 1m, respectively, were constructed and load-tested at one of the two National Geotechnical Experimentation sites located at the campus of Texas A&M University at Riverside. Before footing construction, field and laboratory soil investigations were carried out to obtain soil behavioral information and the results were provided to those interested in making a prediction of the load-settlement behavior of the footings. The predictions were published in a 'prediction volume' which also contains the field and laboratory test results, the footing locations and dimensions, the load testing procedure and results, and a summary analysis of the predicted and observed behavior of the footings.

This paper is a part of the 'prediction volume' and it contains the author's prediction of the following aspects of the behavior of the footings: (i) the loads (Q25) on the five footings corresponding to a settlement of 25 mm at 30 minutes after load application, (ii) the loads (Q150) on the five footings corresponding to a settlement of 150 mm at 30 minutes after load application, (iii) the creep settlements of Footings 1 and 3 due to the corresponding loads Q25 during the period from 1 minute to 30 minutes after load application and (iv) the settlements of Footings 1 and 3 due the corresponding loads Q25 at 20 years after load application. For the sake of brevity, details of the soil investigation program, the soil characteristics or the prediction methodologies are not presented herein. Only an outline of the soil investigation program, and a brief description of the soil parameters and the conceptual basis of the prediction methodologies are presented.

FIELD AND LABORATORY SOIL INVESTIGATION

The field exploration program conducted at the load test site consisted of 6 Standard Penetration Test (SPT) borings, 5 Electric Piezocone Penetrometer Tests (CPT), 4 Pressuremeter Tests (PMT), 2 series of Cross-Hole Wave Tests (CHT), 3 Dilatometer Tests (DMT), 3 Bore Hole Shear Tests (BHST), and a Step Blade Test (SBT) with four blades in one borehole. The laboratory tests performed on the sandy soil samples obtained from the test site included visual soil classification, sieve analyses, and determination of water content, relative density and unit weight. Triaxial compression and resonant

[1] Geotechnical Engineer, Ardaman & Associates, Inc., 8008 South Orange Avenue, Orlando, Florida 32809

column tests were performed on samples of sandy soil reconstructed to the estimated in situ relative density. Since the predictions presented in this paper are based on the results of the SPT, CPT and PMT tests performed at the site, only the soil parameters obtained from these tests are briefly discussed herein.

SOIL CHARACTERISTICS

Based on the results of the field and laboratory soil investigation compiled by Gibbens and Briaud (1993), the soil profile at the test site is comprised of predominantly sandy soil to a depth of about 11 meters. From the ground surface to a depth of about 3.5 meters, the soil is comprised of a medium dense, tan silty fine sand. The SPT blow count (i.e., N-value) of this material ranged from 10 to 30 with an average N-value of approximately 17. The average cone tip resistance from the CPT tests within this depth range was approximately 7,000 kN/m^2 and the average initial modulus from the PMT tests within this depth range was approximately 9,000 kN/m^2. The soil within this depth range had a fines content of less than 25 percent (based on four grain size analyses), a relative density of approximately 55 percent and a friction angle of approximately 35 degrees.

Underlying the silty fine sand layer is a layer of medium dense silty sand with clay and gravel extending to a depth of about 7 meters below ground surface. The N-value for this material ranged between 14 and 34 with an average N-value of approximately 20. The average cone tip resistance from the CPT tests within this depth range was approximately 8,000 kN/m^2 and the average initial modulus from the PMT tests in this depth range was approximately 9,000 kN/m^2. The soil within this depth range had a fines content of less than 10 percent (based on one grain size analysis).

Between the depths of 7 and 11 meters below ground surface, the soil is comprised of medium dense silty sand to sandy clay with gravel. The N-value for this material ranged between 8 and 64 with an average N-value of approximately 30. The average cone tip resistance from the CPT tests within this depth range was approximately 6,000 kN/m^2. Based on the results of one grain size analysis, the soil in this depth range had a fines content of less than 35 percent.

A layer of hard, dark gray clay was encountered between the depths of 11 and 33 meters below ground surface. The N-value for the clay material between the depths of 11 and 15 meters below ground surface ranged between 36 and 99 with an average N-value of approximately 52. The average cone tip resistance from the CPT tests within this depth range was approximately 9,000 kN/m^2. The clay in this depth range had an average moisture content of approximately 28 percent.

Before conducting the field exploration program, surficial soils ranging in thickness from 0.5 to 1.5 meters were removed from the test site. The groundwater table observed in an open well at the site was at a depth of 4.9 meters below ground surface.

FOOTING CHARACTERISTICS

The five spread footings employed in the load tests are designated as Footing 1 through Footing 5 and have approximate dimensions of 3m x 3m, 1.5m x 1.5m, 3m x 3m, 2.5m x 2.5m and 1m x 1m, respectively. Approximate thickness of the five footings are 1.22, 1.22, 1.35, 1.22 and 1.17 meters, respectively, and the footings were installed at approximate depths of 0.76, 0.76, 0.89, 0.76 and 0.71 meters below ground surface, respectively.

PREDICTION OF LOADS AND SETTLEMENTS

The predictions presented herein are based on the results of SPT, CPT and PMT tests, used in conjunction with the settlement prediction methods proposed by Burland and Burbidge (1985), Schmertmann (1970) and Schmertmann et al. (1978), and Menard and Rousseau (1962), respectively.

The method proposed by Burland and Burbidge (1985) is based on a statistical analysis of 200 case histories of settlement of footings, tanks and embankments on sand and gravel. In this method, the settlement of a footing is expressed as a function of the bearing pressure, the width of the loaded area and the average N-value within the depth of influence of the footing. Effects of other influencing factors including depth and shape of foundation, depth to water table and time after load application are also accounted for in this method.

In the method proposed by Schmertmann (1970) and Schmertmann et al. (1978), the settlement is computed by integrating the vertical strain within the depth of influence of the footing. The vertical strain distribution used in this method is based on elastic theory, results of finite element analyses and model studies of shallow foundations. The soil stiffness used in the computation of vertical strain is obtained from a correlation with the cone tip resistance obtained from CPT tests, and correction factors are used to account for the effects of depth of embedment and time after load application.

The method proposed by Menard and Rousseau (1962) is based on a semi-empirical equation in which the settlement of a footing is expressed as a function of the initial modulus obtained from PMT tests, the average bearing pressure, the width, depth and shape of the footing, and an empirical rheological factor that depends on the soil type and the initial modulus to limit pressure ratio obtained from PMT tests.

Loads for 25 mm Settlement

For each footing, the load (Q25) corresponding to a settlement of 25 mm on the 30-minute load-settlement curve was first determined independently using the three methods described above. The loads so obtained were then multiplied by weighing factors and added together to obtain a weighted average load for each footing. The weighing factors were selected based on the relative amount of available data from SPT, CPT and PMT tests, the locations of various tests relative to the footing locations and the author's judgement on the applicability of the corresponding prediction methods. The weighted average loads for the five footings are presented in Table 1. As shown in the table, the loads Q25 for Footings 1 through 5 were estimated to be 3000, 800, 2800, 2000 and 550 kN, respectively.

Loads for 150 mm Settlement

The load-settlement relationships for all five footings at a settlement of 150 mm were determined to be non-linear and, therefore, the prediction methods described above could not be used directly to estimate the loads for a settlement of 150 mm. To determine the loads (Q150) corresponding to a settlement of 150 mm, a load-settlement curve was empirically constructed for each footing using the predicted load-settlement relationships for 25 mm settlement and observed load-settlement behavior of footings of similar size and resting on similar soils. The loads Q150 were then obtained from the empirically constructed load-settlement curves. As shown in Table 1, the loads Q150 for Footings 1 through 5 were estimated to be 15000, 2500, 14000, 9000 and 1100 kN, respectively.

Creep Settlements between 1 and 30 Minutes

The creep settlements of Footings 1 and 3 due to the corresponding loads Q25 during the period from 1 minute to 30 minutes after load application were estimated using the modulus versus time relationships obtained from the PMT tests performed at the site. First, the settlements at 1 minute after load application were computed from those at 30 minutes after load application using the modulus versus time relationships. The creep settlements were then computed as the difference between the settlements at 1 minute and those 30 minutes after load application. As shown in Table 1, the creep settlements for Footings 1 and 3 were estimated to be 1.4 and 1.6 mm, respectively.

Settlements in the year 2014

The settlements of Footings 1 and 3 in the year 2014 (i.e., after 20 years of sustained loading) due to the corresponding loads Q25 were computed using the settlement versus time relationships proposed by Burland and Burbidge (1985) and Schmertmann (1970). Based on the settlements computed using the above two methods, the settlements of Footings 1 and 3 in the year 2014 were estimated to be 32 and 35 mm, respectively, as shown in Table 1.

Table 1. Predicted Loads and Settlements

Parameter	Footing Number and Dimensions				
	Footing 1 3m x 3m	Footing 2 1.5m x 1.5m	Footing 3 3m x 3m	Footing 4 2.5m x 2.5m	Footing 5 1m x 1m
Load (Q25) for 25 mm of settlement on the 30-minute load-settlement curve (kN)	3000	800	2800	2000	550
Load (Q150) for 150 mm of settlement on the 30-minute load-settlement curve (kN)	15000	2500	14000	9000	1100
Creep settlement between 1 minute and 30 minutes due to Q25 (mm)	1.4		1.6		
Settlement in the year 2014 due to Q25 (mm)	32		35		

CONCLUDING REMARKS

The objective of this paper was to see how well the behavior of spread footings on sand could be predicted using results of SPT, CPT and PMT tests. For a comparison of predicted versus observed behavior of the footings, this paper should be read in conjunction with the load test results presented in the proceedings of the symposium.

APPENDIX. REFERENCES

Burland, J. B., and Burbidge, M. C. (1985). "Settlement of Foundations on Sand and Gravel", *Proc. Instn. Civ. Engrs.*, Part 1, 78, 1325-1381.

Gibbens, R., and Briaud, J.-L. (1993). "Data and Prediction Request for the Spread Footing Prediction Event Sponsored by FHWA at the Occasion of the Specialty Conference: SETTLEMENT '94", Dept. of Civ. Engrg., Texas A&M University.

Menard, L., and Rousseau, J. (1962). "L'evaluation des Tassements-Tendances Nouvelles", *Soils-Soils*, 1(1), 13-29.

Schmertmann, J. H. (1970). "Static Cone to Compute Static Settlement over Sand", *J. Soil Mech. Found. Div.*, ASCE, 96(3), 1011-1043.

Schmertmann, J. H., Hartman, J. P., and Brown, P. R. (1978). "Improved Strain Influence Factor Diagrams", *J. Geot. Engrg. Div.*, ASCE, 104(8), 1131-1135.

LOAD-SETTLEMENT CHARACTERISTICS OF SPREAD FOOTINGS

V.A. Diyaljee[1]

INTRODUCTION

In designing spread footings for buildings and other civil engineering structures it is necessary to determine the **ultimate and allowable bearing capacities** and **settlement** characteristics of the foundation soils on which these footings bear.

The purpose of this paper is to provide a prediction of **load and settlement behaviour** at small and large deformations of **five (5)** spread footings founded within a depth of 1m in a sandy soil. The technical information pertaining to the foundation soils, footings, and field and laboratory tests have been provided to **predictors** in a report entitled **'Data and Prediction Request For Spread Footings'** included as an **Appendix** to this prediction volume.

SCOPE OF PREDICTION

The scope of the prediction requires the following to be determined:

(A) For each of the five (5) footings

1. What will be the load measured in the load test at a settlement of 25mm on the 30 minute load-settlement curve.
2. What will be the load measured in the load test at a settlement of 150 mm on the 30-minute load-settlement curve.

[1]Assistant Director, Geotechnical Services Section
Roadway Planning Branch, Alberta Transportation & Utilities,
Alberta, Canada

(B) For the 3m x 3m footing

(3) For the load corresponding to a settlement of 25mm on the 30-minute load-settlement curve, what will be the amount of creep settlement which will take place between the 1-minute and 30-minute reading.

(4) For the load corresponding to the settlement curve, what will be the settlement in year 2014 if that load is applied for the next 20-years.

METHOD OF APPROACH (FIVE FOOTINGS)

Although several insitu test results have been presented, **the conventional soil mechanics approach** utilizing results from laboratory tests and empirical data presented in **Soil Mechanics and Foundation Engineering texts, and Conference papers.** Along with these, **engineering judgement** was used to modify numbers based on **experience and intuitive assessments.** The predictions are summarized in the **Prediction Summary Table** on Page 3.

Load Capacity Determinations

The **ultimate bearing capacity and hence the ultimate or failure load was determined** using **Equation (2), page 168** of the **Canadian Foundation Engineering Manual (1992).**

The **loads** likely to cause **25mm** settlement were derived from **Fig 4.7, page 221, 'Foundation Analysis and Design'** by **Bowles (1988)** with modifications using the **triaxial test results** to represent the behaviour of a **model footing.**

For each footing, the depth at which the soil would undergo significant deformation was considered to be within the assumed triangular wedge shape of the **slip-lines** in the **ultimate bearing capacity condition.** These depths varied from 2.83m for the **3.004m** wide footing to **1m** for the **0.991m** footing.

Using the triaxial test results with curves mid-way between confining pressures of **138 and 345 kN/m^2** provided **axial strains** that were used to check whether the loads that were determined through Fig 4.7 were satisfactory. Slight modifications were made including a decrease in pressure of **20 kN/m^2** for each footing to allow for creep effects from 1 to 30 min since it was assumed that the load would have to be kept constant or be decreased to maintain the 25 mm settlement. The **20kN/m^2** pressure was determined from the triaxial results assuming a **0.15%** rebound strain.

Load Causing 150mm Settlement

For the **150mm movement**, the load that could be operative can be determined from **Equation 4.12, page 221, Bowles (3).** Using this relationship the bearing pressures and hence loads were determined

PREDICTION SUMMARY TABLE

	FOOTING 1 3 m X 3 m (3.004)	FOOTING 2 1.5 m X 1.5 m (1.505)	FOOTING 3 3 m X 3 m (3.023)	Footing 4 2 m X 2 m (2.489)	Footing 5 1 m X 1 m (0.991)
LOAD FOR 25 mm OF SETTLEMENT Q25 ON THE 30 MINUTE LOAD SETTLEMENT CURVE (kN)	2526	788	2552	1801	422
LOAD FOR 150 mm OF SETTLEMENT Q150 ON THE 30 minute LOAD SETTLEMENT CURVE (kN)	7219	1576	7293	5280	613
CREEP SETTLEMENT BETWEEN 1 MINUTE AND 30 MINUTE FOR Q25, s (mm)	6 mm		6 mm		
SETTLEMENT IN THE YEAR 2014 UNDER Q25 (mm)	64 mm		64 mm		

to be in excess of that which would theoretically result in failure of the footings. Since these values exceed the ultimate load capacities it is expected that failure would have occurred at a load much lower than determined from **Equation 4.12.**

For each of these footings, the bearing pressure, and hence load that would be operative, can be determined from the triaxial test curves. For the desired **150mm settlement,** the curves with **345kN/m^2** confining pressure were utilized with slight modifications.

SETTLEMENTS (3mx3m Footing)

Creep Settlement (1 to 30 min)

The creep settlement that is expected to occur has been determined to vary from **0.2mm/hr to .06mm/hr** from paper by **Kamenov (1979).** From these values with modifications for judgement, a value of **6mm** was assessed.

Settlement after 20 years

For this determination it was assumed that the water table would rise at some time to within a depth of the underside of the footing or that **environmental effects** would allow **wetting up of the foundation soils** to cause a **reduction in load capacity.** Hence, it is assumed that the settlement would **double** providing for **50mm** to which **6mm creep** is added plus an additional **8mm creep movement** for the **20 year period**, resulting in **64mm** of predicted movement. The additional 8mm of movement was determined using a creep rate of **0.00005mm/hr.**

CONCLUSIONS

It is expected that results from the actual testing may differ from the predictions provided. The approach used is **quick** and may **lack glamour** in the light of the more **modern approaches** using the results of pressuremeter, dilatometer, and cone penetrometer tests. Use of these tests are not generally the norm unless soil conditions, structure loads, and structure type suggest a **high risk condition.**

REFERENCES

"Shallow Foundations". ***Canadian Foundation Engineering Manual,*** 3rd Edition (1992)., 149-172.

Bowles J.E (1998). ***"Foundation Analysis and Design"***.,4th Edition Chapter 4., 179-239

Kamenov B. (1979). **"Strain rate of sand under one dimensional consolidation".,** *Proc., of the 7th European Conference on Soil Mechanics and Foundation Engineering,* **Vol 2., 67-68.**

Subject Index

Page number refers to first page of paper

Author Index

Page number refers to the first page of paper